Identification Modeling and Characteristics of Miniature Rotorcraft

IDENTIFICATION MODELING AND CHARACTERISTICS OF MINIATURE ROTORCRAFT

BERNARD METTLER

Kluwer Academic Publishers
Boston/Dordrecht/London

Distributors for North, Central and South America:
Kluwer Academic Publishers
101 Philip Drive
Assinippi Park
Norwell, Massachusetts 02061 USA
Telephone (781) 871-6600
Fax (781) 681-9045
E-Mail: kluwer@wkap.com

Distributors for all other countries:
Kluwer Academic Publishers Group
Post Office Box 322
3300 AH Dordrecht, THE NETHERLANDS
Telephone 31 786 576 000
Fax 31 786 576 474
E-Mail: services@wkap.nl

Electronic Services < http://www.wkap.nl>

Library of Congress Cataloging-in-Publication Data
Mettler, Bernard
Identification Modeling and Characteristics of Miniature Rotorcraft
ISBN: 1-4020-7228-7

Printed on acid-free paper.

Printed in Great Britain by IBT Global, London

Contents

Foreword

Things that fly – kites, balloons, blimps, airplanes, and helicopters – seem to fascinate people, especially when they fly by themselves. The robotics field has developed autonomous land vehicles and unmanned underwater robots. The next is, naturally, autonomous flying robots – commonly called unmanned aerial vehicles or UAVs. This book, based mostly on the PhD thesis of 2001 by Bernard Mettler at Carnegie Mellon University, deals with a fundamental problem for autonomously flying a helicopter.

There is a growing interest in developing an unmanned autonomous helicopter, in particular, a small-scale helicopter. A helicopter has unique capabilities – vertical take off and landing from unprepared sites, broad envelope of flight ranging from hovering to cruising, potential to fly at low altitude, and highly agile maneuvering in tightly constrained environments. These unique flight characteristics of the helicopter suggest a wide range of applications in both the military and civilian sectors. In a military role, low-cost unmanned helicopters are being investigated for reconnaissance, urban surveillance, search and rescue, and even for weapon pointing and delivery. The Army/NASA Rotorcraft Division at the Ames Research Center has played an active role in the development of a range of autonomous rotorcraft for military applications from small (e.g., 9 inches diameter) ducted fans configurations (referred to as "Micro Air Vehicles (MAV)" and "Organic Air Vehicles (OAV)") to unmanned versions of full size helicopters. Among the most futuristic and challenging missions is remote navigation, landing, and take-off at remote urban sites under vision-based navigation. In civil applications, low-cost unmanned autonomous helicopters can be dispatched to search for victims of a disaster or to patrol a sensitive area. They can be sacrificed in dangerous conditions – flying close to a forest fire, identifying radioactive leaks, and sampling bio-hazardous materials. Applications expand – exterior inspection of a large man-made construction, wild life observation, stunt cinematography, and so on. Indeed, Carnegie Mellon's experimental 10ft rotor diameter unmanned helicopter was used to map a Martian meteorite impact crater in an Arctic island.

Developing an autonomous unmanned small helicopter presents different and difficult challenges than most of the land-based robots. A helicopter is inherently unstable, high-order, cross-coupled, and its dynamics must be understood and modeled for it to be controlled continuously to function as a stable mobile platform. Bernard's thesis is one of the first significant attempts to precisely model and identify the system characteristics of a small-scale unmanned rotorcraft for advanced flight control design. His approach is to develop a simple but effective linear parameterized model of vehicle dynamics by using a proven system identification tool CIFER® (Comprehensive Identification from Frequency Responses), developed by the Army/NASA Rotorcraft Division. Flight data were collected from Carnegie Mellon's R-50 small-scale helicopter, and the results were verified and used to further optimize its control performance. Bernard's thesis was also the first application of the advanced control system optimization tool CONDUIT® (Control Designer's Unified Interface, developed by the Army/NASA Rotorcraft Division), to systematically and efficiently tune control laws for a model-scale UAV helicopter against multiple and competing dynamic response criteria. This book presents the detailed account of how the theory was developed, the experimentation was performed, and the results were used. It serves as basic and illustrative reading material for all students that are interested in developing autonomous flying helicopters.

Robotics is a science of integration. Development of capable task-worthy robots requires a framework for theories of traditional disciplines and experimentation of system development to interact. For flying robots, system dynamics is where the two meet. Flying robotics is a relatively new area in robotics. In addition to gaining an understanding of the technical contents, we hope that readers will experience from the book the excitement that interdisciplinary activities bring to robotics research.

"Flying robots will bring robotics to new heights "

Prof. Takeo Kanade
U. A. and Helen Whitaker Professor of Computer Science and Robotics
Carnegie Mellon University

Dr. Mark B. Tischler
Flight Control Technology Group Leader
Army/NASA Rotorcraft Division (AMCOM)
Mofett Field, CA

Acknowledgments

Most of the material in this book comes from my Ph.D. dissertation at Carnegie Mellon University. After its completion I joined the Laboratory for Decision and Information Systems at MIT for a postdoctoral residence where I had the opportunity to apply some of my earlier results and give a final shape to this manuscript.

At Carnegie Mellon, I would like to thank my advisor, Prof. Takeo Kanade for his support and guidance during my thesis research. I also owe a great deal to the members of Carnegie Mellon's Autonomous Helicopter Project: Dr. Omead Amidi, Mark DeLouis and Dr. Ryan Miller. I am also very grateful to Dr. Mark Tischler who made his expertise in rotorcraft modeling and identification available to me. I am also thankful to the Army/NASA Rotorcraft division who financially supported me during my doctoral studies (NASA Grant NAG2-1441) and gave me an opportunity to spend two summers in a unique and stimulating environment. I would also like to express my gratitude to Prof. Howard Curtiss for the detailed review he gave to my Ph.D. dissertation, as well as to my other thesis committee members, Prof. William Messner and Prof. Howard Choset. A number of people gave me very constructive feedback on the technical content and writing. Among these individuals, I would like to particularly thank Beth Pillsbury for proofreading the thesis and parts of the new manuscript, Marco La Civita for technical feedback on my thesis, and my good old friend Dr. Zsolt Kukorelly who shortly before turning in my dissertation flew to Pittsburgh to help me port my thesis to a stable and predictable text editing system (from guess what word processor).

At MIT, I would first like to thank Prof. Eric Feron for giving me the opportunity to work with him and his students. I am particularly grateful to Vladislav Gavrilets and the other members of the MIT helicopter team, Ioannis Martinos and Kara Sprague. I would also like to thank Lauren Clark for the thorough proofreading, as well as William Litant for an earlier review of the manuscript. I also learned a lot about communicating ideas by teaching a first course in aerial robotics (Course 16-399) and trying to satisfy the eager curiosity of my students. Finally, transforming my thesis into a book was not straightforward. In this process I received invaluable feedback from Chris Dever (if he were as good a cyclist as a scientific and literary critic, he would have a chance at the Tour or even beating A.Z.). Funding during my postdoctoral residence at MIT came from DARPA/SEC (F33615-01-C-1850), NASA (NAG2-1482), and Barron Associates (264-SC01). I am also very grateful to the Labor and Pax Foundation for their support during my graduate studies.

to my parents and grandparents

Nomenclature

Note that the parameters of the model can be found in Tables 4.3 and 4.4, and the helicopter variables shown in the helicopter reference frame are illustrated in Figures 3.1 and 3.2.

a	longitudinal rotor flapping angle
a_0	blade coning angle
a_n	n-th order harmonic longitudinal flapping coefficient
A	main rotor disc area
$\boldsymbol{A}$	state-space system matrix
A_1	lateral cyclic blade pitch
A_{lon}	longitudinal stick to cyclic pitch gearings
b	lateral rotor flapping angle
b_n	n-th order harmonic lateral flapping coefficient
$\boldsymbol{B}$	state-space input matrix
B_1	longitudinal cyclic blade pitch
B_{lat}	lateral stick to cyclic pitch gearings
β	blade flapping angle
c	longitudinal stabilizer bar flapping angle
$C_{l\alpha}$	blade lift-curve slope
C_T	main rotor thrust coefficient
d	lateral stabilizer bar flapping angle
$\boldsymbol{F}$	total external forces acting on the helicopter center of gravity
G_{uu}	input auto-spectrum function
G_{yu}	cross spectrum
h	distance between rotor hub and fuselage center of gravity
h_{cg}	offset in the center of gravity position
H_{yu}	multivariable frequency response
$\boldsymbol{I}$	helicopter inertial tensor
I_β	blade moment of inertia about the flapping hinge
$J(\Theta)$	frequency domain cost function
k	spring constant of blade flapping spring
K_d	stabilizer bar gearing
k_β	flapping hinge restraint spring constant
K_β	effective rotor spring constant
L_b	lateral rotor moment (flapping spring) derivative
m	helicopter mass
m_b	mass of main rotor blade
$\boldsymbol{M}$	total external moments acting on the helicopter fuselage
M_a	longitudinal rotor moment (flapping spring) derivative
N	scale ratio
N	number of samples in one input-output segment

N_d	number of recorded input-output segments
n_ω	number of fitting frequency points
p	roll rate in helicopter coordinate frame
q	pitch rate in helicopter coordinate frame
r	yaw rate in helicopter coordinate frame
T	main rotor thrust
T_d	length of input-output data segments
T_s	sampling interval of input-output data
T_T	tail rotor thrust
$\boldsymbol{T}$	main rotor thrust vector
$\boldsymbol{T}_{yu}$	mutlivariable transfer function
u	helicopter longitudinal speed in helicopter frame
$\boldsymbol{u}$	vector of system inputs
v	helicopter lateral speed in helicopter frame
$\boldsymbol{v}$	velocity of the helicopter center of gravity
w	helicopter vertical speed in helicopter frame
$\boldsymbol{w}$	helicopter angular rates
$\boldsymbol{x}$	vector of system states
$\boldsymbol{y}$	vector of system outputs
α_D	hub plane angle of attack
β	blade flapping angle
δ_{lat}	cyclic lateral control input
δ_{lon}	cyclic longitudinal control input
δ_{ped}	pedal control input
δ_{col}	collective control input
$\boldsymbol{\varepsilon}$	vector of frequency response magnitude and phase errors
γ	blade Lock number
γ_{yu}^2	coherence function
λ_i	uniform steady-state rotor inflow ratio
λ_β	normalized blade flapping natural frequency
μ	advance ratio
ν_β	undamped natural flapping frequency ratio
Ω	rotor speed
Ω_s	frequency sampling interval
ω_n	discrete frequency points i.e. frequency samples
ϕ	Euler angle for helicopter roll
Φ	rotor inflow angle
ψ	Euler angle for helicopter heading
Ψ	blade azymuth angle
τ_f	main rotor time constant
τ_s	stabilizer bar time constant
θ	Euler angle for helicopter pitch
Θ	blade pitch angle
$\boldsymbol{\Theta}$	vector of unknown model parameters

v_i	uniform steady rotor induced velocity
ω_β	natural blade flapping frequency
$W(\omega_n)$	weighting function for frequency response cost function
X_a	longitudinal rotor force derivative
Y_b	lateral rotor force derivative

Acronyms

CIFER	Comprehensive Identification from FrEquency Responses
PID	proportional-integral-derivative
R-F	coupled rotor-fuselage mode
RUAV	rotorcraft UAV
TPP	rotor tip-path-plane
UAV	unmanned aerial vehicle

Chapter 1

MOTIVATION AND BACKGROUND

1. Overview

The successful development of an autonomous aerial vehicle requires solving complex engineering problems. Toward the end of the 1980's, the shrinking size of computers and the arrival of small commercial global positioning systems (GPS) units and inertial sensor packages made it possible to instrument a small-scale rotorcraft for free and automatic flight. In these early years of aerial robotics a small number of these systems were built, mainly in academic institutions, and a few were able to demonstrate basic flight capabilities such as hover at a position and slow flight following waypoints. In the period following these important breakthroughs, little progress were made in improving the automatic flying capability; the results were far from the theoretical potential of the vehicle. The main reason for this limitation was the absence of an accurate model of the vehicle dynamics that could be used for the analysis and design of the flight control system. These early results were often obtained using classical control architectures with empirically tuned feedback gains.

The dynamics of rotorcraft are significantly more complex than those of a fixed wing aircraft. Without a good understanding of their dynamics, the operational capabilities of aerial robots based on small-scale helicopters will remain limited. Using models developed from first-principles alone did not fulfill the needs mainly because there was no real sense of the complexity and hence of the degree of simplification that could be reasonably allowed. Without knowledge of key dynamic characteristitcs of these vehicles it was impossible to arrive at a model that was accurate yet simple enough to be practical for control analysis and design.

Interested by the application of modern control design techniques to small-scale rotorcraft, I needed such a model. System identification had played an important role in the development of flight control systems for full-scale rotorcraft. So I undertook the project of applying state-of-the-art rotorcraft system identification techniques to Carnegie Mellon University's (CMU) instrumented Yamaha R-50 helicopter. This book describes in detail the modeling and identification process, the analysis and characterization of the dynamics of miniature rotorcraft, and the analysis of the control system originally used on CMU's helicopter; using both the model and closed-loop identification techniques. The results from the control system analysis explain why the performance of a classical control architectures is inherently limited. Since these modeling results were originally published, important milestones were obtained, signaling a clear take-off of the aerial robotics field.

In this chapter, I first briefly detail the motivation for using rotorcraft-based unmanned aerial vehicles. Starting with an overview of the early developments in aerial robots based on small-scale rotorcraft, I highlight the key technical and theoretical challenges, the solutions employed and their limitations, and the requirements for a successful development. Section 2 describes the flight test vehicle used in the original work (Carnegie Mellon's Yamaha R-50) and the vehicle used to illustrate some of the key characteristics of small-scale rotorcraft (MIT's X-Cell .60). In section 3, I give a technical background to the rotorcraft modeling and control field and its application to small-scale rotorcraft. Finally, in section 4, I state the original objective of my identification modeling research and give a preview of the material and results presented in the book. The material in this chapter was not updated for the progress made since the completion of my PhD dissertation in December 2000. Therefore it gives a good sense of the state of the field when I started the project. The last book chapter will give an overview of the key milestones that have been obtained since then.

1.1 Motivation

Unmanned aerial vehicles (UAVs) offer alternatives to manned aircraft for missions that can be dangerous for crewmembers, where automation can improve efficiency, or where applications are impossible for manned aircraft due to other constraints such as the necessity of a miniaturized vehicles. Fixed-wing UAVs are already used for applications such as high atmosphere research (e.g. Helios, Pathfinder) and or battleground monitoring (e.g. Predator, Global Hawk).

With the ability to take off and land vertically and hover, along with natural agility and controllability, rotorcraft will extend the potential roles for UAVs. Helicopters already have an irreplaceable role among aircraft, indispensable for a variety of tasks ranging from medical evacuation to transportation to construction in confined areas. Such rotorcraft UAVs (RUAV) are already greatly valued by the military for applications for a range of battlefield tasks, such as exploration and even combat operations. There are also numerous examples for civilian applications including filmmaking (allowing both steady and dynamic aerial views), close-up inspection (bridges, buildings, dams), and digital terrain modeling (where a small vehicle, because of its potential for closer proximity to the terrain and structures, could gather more detailed features). One could also imagine applications where an RUAV would manipulate its immediate environment, depositing or picking up objects or probing the soil. There are even plans for using rotorcraft for the exploration of planets like Mars [16, 45]. Such a flying exploration robot is certainly attractive considering the difficulties involved in driving through rough terrain. Small-size rotorcraft tend to be naturally more maneuverable and more responsive than traditional full-scale rotorcraft. The maneuverability of such vehicles can offer a tremendous operational advantage if harnessed during autonomous flight. Taking full advantage of the small-scale helicopter's natural abilities, however, has been a major challenge to its use as a UAV platform.

Simply put, helicopters are difficult to fly. They are unstable, requiring continuous attention from the pilot. Four primary inputs are necessary to control their motions: the longitudinal and lateral cyclic for horizontal motions in their respective directions; collective for vertical motion; and pedal for yaw (heading) motion. The response of the vehicle to each of these inputs has secondary effects in addition to the primary, intended response. Therefore, to obtain a precise vehicle response, the pilot needs to compensate the secondary responses by simultaneous action of all inputs. Helicopters is a good example of a multivariable system. Moreover, as the vehicle changes flight conditions, its dynamic characteristics change; the vehicle responses to control actions in hover flight differ from their responses in cruise flight. In hover, it is free to move in any of three spatial dimensions and even rotate about its vertical axis, while in cruise, the helicopter behaves very much like a fixed-wing aircraft. Helicopters are also sensitive to atmospheric disturbances, such as wind gusts or turbulence. This sensitivity can cause difficulties when the helicopter operates near buildings, structures, or other moving vehicles. However, compared to fixed wings, rotorcraft can quickly produce large control forces, therefore, with high-bandwidth control (or skilled pilot) a rotorcraft can tolerate gusty conditions.

More so than for other mobile robots, the development of a working aerial robot requires the successful integration of several disciplines. The tasks and issues involved with aerial robots tend to be more complex, and failures can easily result in the destruction of the vehicle. These challenges, however, ultimately make aerial robotics an interesting and prolific research field.

1.2 Evolution of Small-Scale Rotorcraft UAVs

Research projects using small-scale helicopters started in academic institutions in the early 1990's. The vehicles are mostly adapted from helicopters used by hobby radio control (RC) pilots. Such hobby helicopters are inexpensive compared to custom-developed airframes and are available in a broad range of sizes and payloads (from about 1 to 3 meters in rotor diameter with payloads ranging from less than one pound to about 10-20 pounds). The payload is necessary to carry the on-board instrumentation, but can be important for additional equipment such as vision systems. To make a commercial small-scale helicopter a useful experimental platform, the aircraft needs to be equipped with navigation sensors and a flight computer. These systems typically have to be added by researchers.

Early small-scale helicopter projects often used test rigs that prevented the vehicle from crashing and at the same time provided a direct mechanical means to determine the vehicle states (via the joint angles) without requiring on-board navigation sensors. For example, in [94], the helicopter was rigged to a delta robot structure that provided a safety restraint and allowed the helicopter to move about its six degrees of freedom. A similar setup was developed at Carnegie Mellon for the development of flight control systems and vision-based feedback systems [3]. Unfortunately, rigs often limit the number of degrees of freedom and restrict the vehicle to hover flight. Nevertheless, these rigs played an important role in the development of the basic flight control capabilities [93].

Equipping a small-scale helicopter with motion sensors for free flight is difficult due to limited payloads yet stringent navigation requirements for a vehicle capable of simultaneously rotating and translating in all six degrees of freedom. There still exists no commercial single sensor that can directly and accurately measure the rapidly changing vehicle state. Multiple sensors are needed, and their measurements must be integrated to produce accurate estimates of the vehicle states. This state estimation process and the simultaneous flight control process must operate in real time, putting hard requirements on the computing system.

These hardware and software issues fully occupied researchers' time and were a primary reason for the slow and often unsuccessful development of experimental vehicles in the 1990s. Today, these difficulties still represent a major hurdle; however, the situation is improving with accumulated experience and steady progress in sensor and computer technology. This advancement will allow researchers to focus on new topics – such as vehicle dynamics, advanced flight control, and guidance – that are fundamental in creating highly capable vehicles.

Beyond hardware difficulties, fundamental issues remain. Currently, there are no examples of RUAVs that can perform as well under computer control as under human-piloted control. Throughout the 1990s, most RUAVs that demonstrated autonomous flight used simple classical control systems, like the single-loop classical feedback controller used on Carnegie Mellon's [63, 3] or MIT's [19] autonomous helicopters. In these systems, the controller parameters are typically derived empirically (usually hand-tuned). These approaches to flight control cannot account for some critical dynamic effects and thus are inherently limited in the performance they can deliver. The operation of empirically designed controllers is constrained to slow airspeed and small-amplitude maneuvers. Some early systems also used linear quadratic (LQ) controller designs [4], but did not present any advantage due to the limited accuracy of the dynamic models.

The flight control systems of traditional rotorcraft consist primarily of stability augmentation systems (e.g. increasing yaw, roll, or pitch angular damping), and there is little experience with autonomous flight control allowing automatic trajectory following. A number of available control design methods could, in theory, be applied to address the rotorcraft control issues (described later in this chapter). To be successfully applied, these methods require accurate vehicle dynamics models. Since most rotorcraft share the same well-understood physical principles, the available literature on rotorcraft dynamics represents a valid resource for the smaller vehicles.

1.3 Rotorcraft Modeling Challenges

Rotorcraft dynamics are complex, and the development of a model that is at once sufficiently accurate and simple enough for practical control design is challenging.

One major source of complexity is the rotor. The rotor is used to produce the lifting force as well as control forces and torques. Unlike the wing of a fixed-wing aircraft, the rotor is a dynamic system coupled to the vehicle's rigid body dynamics. The rotor response and the resulting forces and torques produced on the fuselage are both functions of the pilot's inputs and the vehicle motion. The control forces and moments are not instantaneous, and building the input action requires attentive observation of the vehicle's transient response. A rigid-body stability derivative model, such as that typically used for fixed-wing aircraft, does not capture the rotor dynamics and hence is not adequate for rotorcraft [33]. For high-bandwidth control, the rotor-fuselage coupling effect must be accounted for in the control system design model [85].

When rotorcraft operate between different flight conditions (e.g. hover, cruise, ascending, and descending flight), the aerodynamic forces acting on the vehicle and the working state of the rotor system change dramatically. This affects the performance of the vehicle as well as how it responds to controls; a rotorcraft behaves very differently in hover flight than it does in cruise flight. This naturally complicates the modeling and control design task.

Rotorcraft Modeling Techniques

The physical approach to modeling involves deriving the equations of motion from the ground up, using the fundamental laws of mechanics and aerodynamics. This approach is referred to as first-principles modeling. For a system involving numerous physical effects, the resulting equations of motion are typically high-order nonlinear coupled vector differential equations. A first principles approach requires considerable knowledge of and experience with all the phenomena involved in rotorcraft flight. However detailed the modeling effort, comprehensive validations against flight data as well as subsequent model refinements are still necessary before the model can be considered accurate. First-principles models are most useful for simulation applications, because they describe the dynamics of the vehicle in a large portion of the vehicle's flight conditions.

For control design applications, as well as for flying qualities studies, simpler, linear models are often sufficient. Linear models have been used extensively and successfully for rotorcraft. At a precise operating point, and even within a certain region around that point, linear models accurately capture the essential effects of the vehicle dynamics. This is advantageous, since numerous analysis and design techniques are available for linear systems. However, to cover the entire flight envelope, multiple linear models and therefore multiple controllers are necessary. An effective approach to deriving accurate linear models of a plant is linear system identification.

System identification and first-principles modeling are ultimately complementary. Applying both approaches is necessary to develop simulation and control design tools, and to tackle the control issues that arise during operation throughout the entire flight envelope (e.g., transition from hover to forward flight).

The Role of System-Identification Modeling

System identification uses experimental input-output data collected from a plant to produce a mathematical representation of the system's dynamics. This method is attractive because it is direct, is based on real data, and integrates validation and model refinement into the modeling process.

In the 1980s, rotorcraft experts undertook important efforts in developing effective identification methods and tools. CIFER (Comprehensive Identification from FrEquency Responses [87]) from the Army/NASA Rotorcraft Division, is one of today's standard tools for rotorcraft identification.

Advanced high-bandwidth flight control design requires models that are accurate at higher frequencies [85]. A very effective and accurate way to obtain linear rotorcraft models is through linear frequency-domain system identification. In this method, frequency responses are estimated from the collected input-output data. Then, the parameters of a linear model capturing the key vehicle dynamics are identified by tuning them to fit the estimated frequency responses.

The identified model can be readily used for control design. In the present situation, system identification can also help recognize the fundamental characteristics of a vehicle. It is hence well suited as a precursor to the more comprehensive first-principles modeling approach. Once accurate linear models have been identified, the information about the dynamic characteristics obtained throughout the identification process can help the derivation of comprehensive first-principles models for small-scale rotorcraft. Finally, system identification can also be conducted on the closed-loop system for an accurate validation and analysis of the controller's performance.

By allowing a tight integration among modeling, simulation, control design and flight-test validation, system identification will play a major role in the development of future aerial robotics systems and other autonomous systems with complex dynamics.

2. Description of the Flight-Test Vehicles

Most results and modeling techniques presented in this book were obtained and developed using Carnegie Mellon's (CMU) Yamaha R-50 helicopter. Later, flight-data was collected from the Massachusetts Institute of Technology's (MIT) X-Cell .60 helicopter, and the approach developed for the R-50 was also applied to that vehicle.

The Yamaha R-50 is a commercially available small-scale helicopter originally designed for remotely operated crop-dusting. Figure 1.1 shows CMU's instrumented R-50, and Figure 1.2 shows its dimensions. The X-Cell .60 is a smaller helicopter that is very popular among hobby pilots for aerobatics and is now being used for research in highly maneuverable autonomous flight. Figure 1.3 shows MIT's instrumented X-Cell; Table 1.1 gives general physical characteristics of the two flight-test vehicles.

The R-50 and the X-Cell helicopters are representative of the two main types of vehicles used for today's research in autonomous small-scale rotorcraft. The R-50 has a large payload (20 kg) and reliable operation, allowing larger sensors (laser range scanners, multiple video cameras) and providing more flexibility with equipment. The X-Cell is highly maneuverable; when not overloaded, it is capable of performing aerobatic maneuvers, and hence is better-suited for research in aggressive flight control and guidance algorithms.

Using two flight-test vehicles of considerably different characteristics allows us to test the general validity of the developed parameterized model and identification approach. At the same time, it helps to illustrate effects of scaling and design on the dynamic characteristics of such miniature rotorcraft.

2.1 Carnegie Mellon's Yamaha R-50

In 1995, CMU equipped a Yamaha R-50 helicopter with custom instrumentation to perform research in vision-based autonomous flight [3]. Later, CMU's RUAV instrumentation was enhanced with an inertial navigation system (INS), which provided accurate measurements of the aircraft's accelerations and angu-

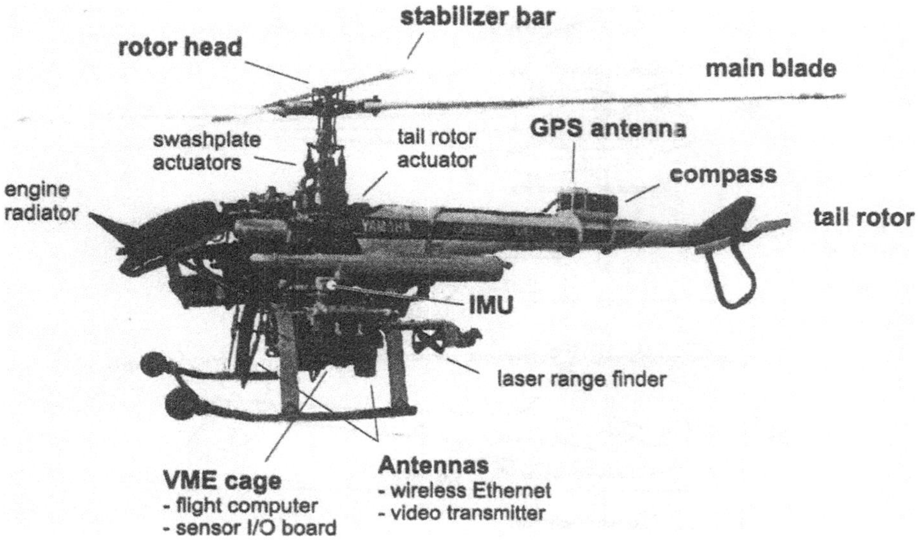

Figure 1.1. Carnegie Mellon's instrumented Yamaha R-50 Helicopter

lar rates; a high-precision GPS receiver, which provided position and velocity information; and a compass for heading. The different sensors were integrated to produce a state-estimation system. With this system, highly accurate estimates of the vehicle states and pilot control inputs can be collected, making CMU's RUAV a perfect platform for the application of system identification.

Airframe Characteristics

The R-50 uses a two-bladed main rotor with a Bell-Hiller stabilizer bar. The relatively rigid blades are connected to a yoke by individual flapping hinges and elastomeric fittings. The yoke itself is attached to the rotor shaft over a teetering hinge in an under-slung configuration (tri-hinge hub; see Figure 3.4). This configuration reduces the amount of radial displacement in the blade center of gravity during blade flapping, which in turn reduces the Coriolis forces that are responsible for a lead-lag blade motion (in-plane blade motion). The teetering motion is also restrained by an elastomer damper/spring. This rotor system

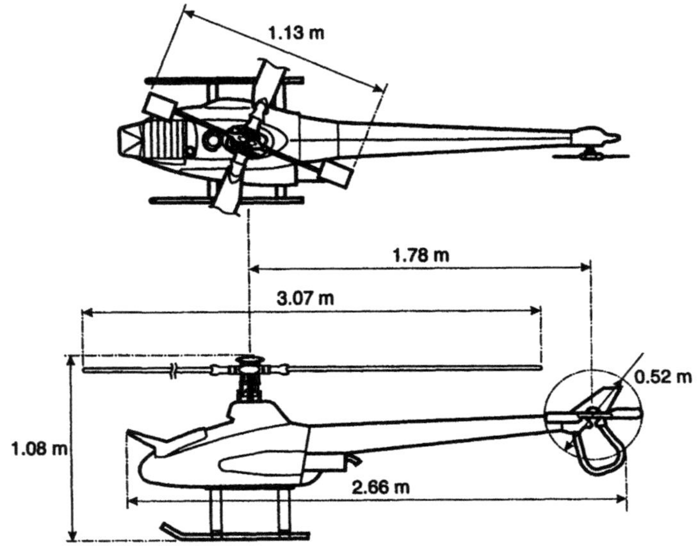

Figure 1.2. Dimensions of the Yamaha R-50 Helicopter

Table 1.1. Physical Characteristics of the Yamaha R-50 and the X-Cell .60

	R-50	X-Cell
Rotor speed	850 rpm	1600-1700 rpm
Tip speed	449 ft/s	419-445 ft/s
Dry weight	97 lb	10 lb
Instrumented	150 lb	17 lb
Engine	2-stroke, water cooled	2-stroke, air cooled
Type of payload	video cameras 3-D laser mapping system	none to date
Flight autonomy	30 minutes	12 minutes

differs from classical teetering rotors in that the teetering motion is restrained and the yoke has separate flapping hinges.

The Bell-Hiller stabilizer bar is a secondary rotor consisting of a pair of paddles connected to the rotor shaft by an unrestrained teetering hinge. The bar receives the same cyclic control input as the main rotor but has a slower response than the main blades. It is also less sensitive to airspeed and wind gusts due to a smaller blade Lock number (non-dimensional rotor parameter giving the ratio between aerodynamic and inertial forces). The stabilizer bar flapping motion is used to generate a control augmentation to the main rotor cyclic input. This augmentation is implemented by the Bell mixing mechanism. From a control point of view, this can be interpreted as a lagged rate (or "pseudo-attitude") feedback in the pitch and roll loops [34, 35]. The low-frequency dynamics are stabilized, which substantially increases the phase margin for the pilot/vehicle system in the crossover frequency range (1-3 rad/sec). The pseudo-attitude feedback also reduces the aircraft's sensitivity to wind gusts and turbulence.

Instrumentation

Carnegie Mellon's (CMU) autonomous helicopter has state-of-the-art instrumentation capable of producing high-quality measurements of the vehicle state variables. The centerpiece of the helicopter on-board systems is a VME-based flight computer, which hosts a Motorola 68060 processor board and a sensor I/O board. All helicopter sensors and actuators connect through the I/O board with the exception of the inertial measurement unit (IMU), which connects directly to the processor board through a special serial port. The flight computer runs under a VxWorks real-time operating system. The communication to the ground station takes place via wireless Ethernet.

CMU's R-50 uses three motion sensors:

- fiber-optic IMU, which provides measurements of the airframe accelerations a_x, a_y, a_z, and angular rates p, q, r (resolution: 0.002 g and 0.0027 deg/sec; data rate: 400 Hz)

- dual-frequency differential global positioning system, or GPS (precision: 2 cm, update rate: 4 Hz)

- magnetic compass for heading information (resolution: 0.5 degrees; update rate: 2 Hz)

The IMU is mounted on the side of the aircraft, and the GPS and compass are mounted on the tail. Velocity and acceleration measurements are corrected for the position offset between the helicopter center of gravity and the sensors. A Kalman filter running at 100 Hz is used to integrate the measurements from the

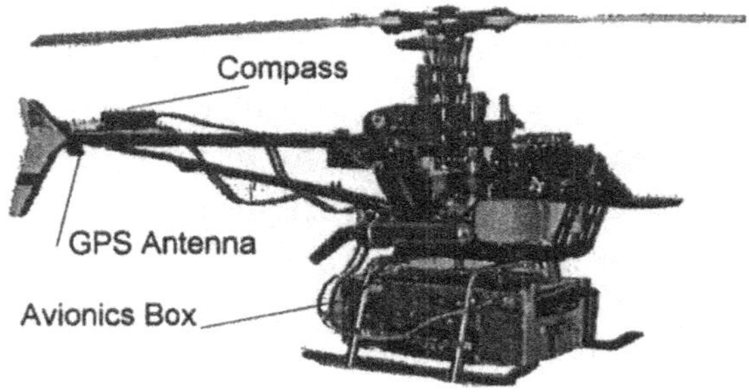

Figure 1.3. MIT's Instrumented X-Cell .60 Helicopter

IMU, GPS, and compass to produce accurate estimates of helicopter position, velocity, and attitude.

Three linear servo-actuators are used to move the swashplate, while another actuator controls the pitch of the tail rotor. The engine speed is controlled by a governor that maintains a constant rotor speed in the face of changing rotor load. The actuators and governor are standard equipment of the R-50.

2.2 MIT's X-Cell .60 Helicopter

MIT's X-Cell helicopter was developed in 1999 as a test vehicle for aggressive maneuvering. The airframe is from an off-the-shelf hobby helicopter popular for its extremely agile flying qualities. The helicopter is currently used for research in computer-controlled aerobatic flight [25, 27, 28] and motion planning for highly maneuverable vehicles [62] (see Chapter 7).

Airframe Characteristics

Apart from size, the principal distinguishing characteristic of the X-Cell is its rigid, hingeless rotor head (see Figure 3.4). This rotor head allows for larger control moments, resulting in higher roll and pitch angular rates. Compared to the R-50, the X-Cell also has a larger thrust-to-weight ratio (between two and three compared to just about 1 for CMU's R-50) permitting fast acceleration and high load factors for tight turns. The rigid rotor head, together with the large thrust, allows for inverted flight. The X-Cell is also equipped with a yaw-damping system and a stabilizer bar. These characteristics, together with the smaller size and weight, combine to make the X-Cell a highly agile vehicle. The differences between the R-50 and the X-Cell will be explained in more detail in Chapter 5.

Instrumentation

For the X-Cell, a great deal of effort was put into the development of instrumentation sufficiently lightweight and compact to preserve the agility of the vehicle. The MIT helicopter features an avionics box that houses all the sensors (except the GPS antenna and the compass), the flight computer, and the batteries. Particular attention was given to the box's vibration isolation, which is a critical problem in rotorcraft. In addition to the typical suite of navigation sensors (i.e. IMU, GPS, compass), the system also includes a barometric sensor. MIT's X-Cell uses a lower-grade and less expensive IMU and GPS than CMU's R-50. A comprehensive description of the system is given in [80].

3. Technical Background

Small-scale rotorcraft have a physical configuration similar to full-scale rotorcraft. Therefore, the techniques developed and used for modeling and controlling full-scale rotorcraft are relevant to small-scale rotorcraft. In the following, we give an overview of the available modeling and control-design techniques, and through examples from literature, we describe how these techniques have been applied to small-scale rotorcraft.

3.1 Rotorcraft Modeling

Two modeling approaches are used for rotorcraft: first-principles modeling (physical modeling), where the vehicle dynamics are modeled from the ground up using the laws of mechanics and aerodynamics [42, 68, 6]; and modeling using system identification, which uses experimental data collected from the vehicle and requires special computational tools.

The first-principles approach is characterized by its comprehensive treatment of all the vehicle's physical features. The goal generally is a model that is valid over a wide range of operating conditions (hover to forward flight), making it suitable for flight simulation applications. Typically, the resulting mathematical equations are nonlinear and have a large number of states. Because of the number of assumptions and simplifications performed during the modeling, the accuracy of such models is not guaranteed. This issue can be addressed by a succession of validation and refinement steps.

First-principles rotorcraft models typically entail dozens of unknown physical parameters that need to be determined. They include the geometrical data about the different components and various physical constants, including: the mass and moments of inertia of the fuselage and rotor components; aerodynamic parameters, such as the blade and stabilizer bar lift curve slope; and the fuselage drag coefficients. The most basic parameters can easily be determined by direct measurements or simple experiments, such as pendulum experiments for the determination of the moments of inertia. Aerodynamic parameters, such as the drag coefficients, which are important for forward flight conditions, usually require more elaborate experiments, such as wind-tunnel testing.

First-principles rotorcraft equations are large and complex vector differential equations that are difficult or even impossible to intuitively understand. More compact and easier-to-understand models that capture the essence of the plant dynamics can facilitate the development and integration of control and guidance systems. System identification has been used to obtain such simple, linear low-order models. The focus of system identification is the development of models that have the right level of accuracy and complexity required by the application. The rotorcraft community has mostly used this approach for flight control design and handling-qualities studies. Compared with first-principles-based approaches, modeling using identification techniques also has the advantage of being based on experimental data. It thus provides real information about the actual vehicle and implicitly includes the model validation and refinement steps. Because of the linear form, however, these models are valid only in the vicinity of an operating point (e.g., hover to slow-speed flight, forward flight, fast-forward flight). As pointed out earlier, system identification and first-principle modeling are ultimately complementary. First-principle modeling provides the necessary understanding of the physical effects involved in the vehicle's dynamics, and system identification provides experimental evidence of these effects and also provides a means to determine physical parameters that cannot be measured directly.

First-Principles Modeling of Small-Scale Rotorcraft

First-principle modeling is the natural approach to modeling when physical understanding is most important. Most researchers working with small-scale rotorcraft have tried this approach for the development of models that they could use for control design. Unfortunately, the experience is often disappointing, because it is tedious and generally does not produce highly accurate results unless performed with extreme care. Weilenmann's thesis [92] is a rare example in which first-principle modeling was used successfully (however, for limited conditions). In his thesis, the author used a rigged small-scale helicopter as a control-system test bench. An accurate model of the helicopter dynamics was needed to apply multivariable control-design techniques. The dynamic model was developed from full-scale helicopter theory. Using this approach, a model in the form of a nonlinear differential equation was obtained. The model was subsequently linearized about the hover operating point to derive a model suitable for linear control theory. This modeling effort concentrated on the hover flight condition, allowing numerous simplifications to be made. The hover model still had 18 states and about 70 parameters. Some of these parameters were directly measurable; however, other parameters needed to be determined experimentally. The modeling of general flight conditions would have been even more complex, and the results would most likely not have been as successful.

Several existing mathematical helicopter models can also be used. Good examples include the Minimum-Complexity Helicopter Simulation Math Model [37] or other models of similar levels of complexity [11, 83, 38]. Such models account for a basic level of complexity and could be modified or used as a template for a different helicopter. However, once such a model (or any first-principle model) has been set up or developed, significant effort is necessary to validate it and subsequently correct the potential shortcomings. Validation requires flight data to be collected. In order to collect flight data, the RUAV has to be instrumented with navigation sensors and a state estimation system. This means that you cannot go around using experimental data for rotorcraft modeling. Once you have the ability to collect accurate data from a vehicle, an identification-based modeling approach becomes possible.

System Identification Modeling of Small-Scale Rotorcraft

System identification is an effective modeling technique used in a number of areas. The difficulties obtaining accurate and practical rotorcraft models for control-design application from first-principles modeling approaches originally led to interest in system identification [31]. In the 1980s, the Army-NASA Rotorcraft Division devised a frequency-domain identification approach and developed the necessary numerical tools [86]. The resulting "Comprehensive Identification from FrEquency Responses" (CIFER) procedure has been applied to numerous full-scale rotorcraft, including the Black Hawk [23], the BO-105 [87], and the SH-2G [89], as well as other aircraft, such as the Harrier and the tilt-rotor XV-15 [84].

System identification allows the characterization of the open-loop aircraft dynamics as well as the characterization of individual components of the closed-loop system from experimental data. This capability makes it an attractive approach for the modeling of systems that combine components of different natures (e.g. mechanical, electro-mechanical, and digital) and systems that are dynamically augmented. The way system identification works – by tapping directly into the system's "information flow" – makes it a natural tool for the integration of the different design steps, starting with the modeling, the performance analysis, and the control design and ending with the flight-test validation. Unmanned aerial vehicles are perfect examples of systems blending mechanical components (e.g. airframe, rotor mechanization), electrical components (e.g. actuators and sensors), and digital components (e.g. flight computers, micro-controllers, sensors). The performance requirements for aerial vehicles are complex, because flight occurs in many degrees of freedom and operational conditions. System identification was recognized as a key tool for the integration of modern aircraft [86] and will play an equally important role in the development of unmanned aerial vehicles.

Frequency-domain identification is better suited than time-domain approaches for rotorcraft identification [84] (see Chapter 2). The first step in a frequency-domain techniques is the estimation of frequency responses from flight data collected from an instrumented flight-test vehicle during special experiments conducted in specific flight conditions. This nonparametric step is valuable in that it provides broad insight into the vehicle dynamics, helping the modeler determine the particular characteristics of the vehicle under investigation. This insight helps the next step of the identification procedure: the development of a parameterized dynamic model. The model is usually a linear state-space model developed from a combination of first-principle knowledge, physical insight, and frequency-response analysis. Subsequently, the model parameters are identified by best matching the model-predicted frequency responses to the estimated ones.

The literature provides only a few examples where system identification techniques were applied to the modeling of small-scale rotorcraft. The following three examples are representative of the early results obtained applying system identification to such vehicles.

In the Caltech experiment [65], the authors used a model helicopter as a test bed for advanced linear and nonlinear control methodologies. To simplify their work, they limited their study to the helicopter's angular motions (roll, pitch, and yaw) and installed their Kyosho EP Concept 30 helicopter on a stand that allowed only angular motion. For the data collection, the authors used small-input perturbations superimposed on the trim-control signal. They chose this approach to make sure that the helicopter response remained in the linear range of the dynamics. The flight data was band-pass filtered to remove the feedback effects of the trim. They used the rigid-body equations of motion as a parameterized identification model. The actuators were included in the model as first-order transfer functions. The parameters of the state-space model were subsequently identified using a discrete-time version of the prediction error method (PEM) [54]. The model was successfully used for model-based control design. The results of the time-domain validation of the model are difficult to interpret because of the band-pass filtering performed on the data and the time scale chosen for the results (40 seconds long runs). For example, the time histories do not show signs of the typical rotor-fuselage coupling, which explains why the rigid-body model results in only a fair match of the experimental data. Moreover, both the open-loop and the closed-loop validations were performed at small amplitudes (± 5 deg) and hence only exploit a narrow range of the vehicle dynamics.

In another example [7], the authors wanted to develop a helicopter simulation model for the evaluation of different control strategies. They proceeded with the identification of the longitudinal dynamics of a Kalt Omega Pro model-scale helicopter using a maximum likelihood technique [54]. To simplify their work, they concentrated on the longitudinal dynamics in hover flight conditions. This simplification was achieved by having the pilot maintain the helicopter in a longitudinal flight configuration during the collection of flight data. In contrast to the Caltech experiment, the control inputs used for the data collection consisted of frequency sweeps (harmonic excitations of increasing frequencies). They also used a rigid-body model for the parameterized identification model but replaced the pitch-damping derivative with a second-order term. This extension of the model was based on the observation that the angular response had a second-order characteristic that the authors attributed to the presence of a stabilizer bar. The model validation results presented in the paper show a relatively good match between the flight data and the responses predicted by the model. Compared to the Caltech experiment, larger inputs were used;

angular responses of up to 100 deg/sec were attained. Also, the second-order dynamics used to augment the rigid-body dynamics contributed to the good match of the pitch dynamics.

In a last example [47], the authors were using a small-scale helicopter as a test bed for nonlinear control and autonomous path planning. They developed a parameterized model using first principles and explicitly accounted for the stabilizer bar. For the identification, they concentrated on the single-input single-output roll and pitch transfer functions. Instead of using a linearized version of the mathematical model for the identification, they used fourth-order transfer functions. They identified the coefficients of the numerator and denominator by minimizing the least squares error between the roll and pitch time-response predicted by the transfer function and the experimental response. The transfer function coefficients were not related to the physical parameters of the mathematical model. Based on the time histories shown in the paper, the match is poor. The responses predicted by the model did not track the faster dynamics of the system. As in the Caltech experiment, the long time histories preclude a detailed inspection of the match.

The results of system identification for the modeling of small-scale helicopters are limited when compared with the results attained for full-scale rotorcraft identification. Small-scale modeling efforts often concentrate on a subset of the vehicle degrees of freedom, the accuracy of the models is poor in frequencies beyond the rigid-body dynamics, and only hover conditions are modeled. The limited results are related to both technical difficulties – illustrating how difficult it is to have a well-instrumented flight-test vehicle – and an often limited understanding of rotorcraft dynamics. Only one experiment was conducted using a free-flying helicopter. By rigging the helicopter, the dynamics of the vehicle can be isolated, allowing one axis to be studied at a time with simple identification tools. Most published research makes little use of the available literature on rotorcraft dynamics and system identification. For example, no model explicitly accounts for the rotor dynamics; rigid-body models are most often used. In the two latter examples, the models were extended to account for the higher-order effects that result from the coupling of the fuselage with the rotor/stabilizer system. However, the results are not physically meaningful. Finally, all identification experiments were conducted in the time domain, and few experiments were performed in a multivariable setup.

Suitability of the Linear Dynamics Assumptions

As with all almost physical systems, rotorcraft dynamics are nonlinear. Therefore, we must understand in what circumstances a linear modeling and control design approach will be adequate to address the main control challenges.

The nonlinearities in rotorcraft arise mainly from two origins: first, the nonlinear kinematics and dynamics associated with the rigid-body equations of motion and gyroscopic forces arising from the spinning rotor; and second, the nonlinear effects inherent to rotorcraft aerodynamics. Around a fixed operating point (defined here in terms of airspeed), nonlinearities are mild, and the vehicle dynamics can be approximated by linear equations of motion. On the other hand, the nonlinearities arising from the changing aerodynamics over the range of useful operating conditions are significant and preclude using the same linear model at different flight conditions. In mathematical terms, the curvature of the vector field in the equations of motion $\dot{\boldsymbol{x}} = \boldsymbol{F}(\boldsymbol{x}, \boldsymbol{u})$ is sufficiently small in the vicinity of most operating points that a linear approximation to $\boldsymbol{F}(\boldsymbol{x}, \boldsymbol{u})$ can be used locally but not over a larger range of conditions. For example, a hover model will not work well in cruise flight, and vice-versa.

Linear approximations of the dynamics have been used for rotorcraft modeling (usually with system identification), handling-qualities evaluations [68, 30], and control design. The stability-derivatives model is the standard linear model used extensively for both fixed-wing aircraft and rotorcraft. To overcome the limitations of linear equations of motion, aircraft dynamics are modeled for different operating conditions (airspeed, altitude, weight). Heffley's report [34] provides the stability-derivative model of several helicopters for various operating conditions. Other examples of such model collections are available in rotorcraft literature [68, 74].

One of the engineering challenges with flight vehicles is controlling the system across the entire flight envelope. Because a linear model will be accurate only within a certain range of the nominal operating point, no global solution to the issue of changing dynamics can be achieved using a linear model. It is important, however, to note that the range of validity for these linear models is usually large enough to be useful for control design. This fact is demonstrated by the large number of linear controllers used for rotorcraft flight control [86, 12, 82]. A single controller, for example, is usually good enough for a range of conditions, such as transition from hover to low speed, or for forward flight at different airspeeds. This characteristic makes linear modeling and linear control a practical approach.

A global solution to the changing dynamics is a goal of nonlinear control. Dynamic inversion and other nonlinear control techniques are attractive for aerospace problems in general because of the wide range of flight conditions encountered. The potential of these techniques has been demonstrated in theory [79, 48]. For many systems with complex dynamics, such as rotorcraft, basic issues of applicability still need to be resolved. For example, in studies of application of nonlinear control to rotorcraft, simplified models of the dynamics are often used. For example, in [48] the authors assume that the helicopter dynamics can be approximated by the rigid-body dynamics and that the coupling between the forces and the moments can be neglected. Such simplifications have important implications and will require more careful study.

Linear models, when combined with other techniques, can be very effective for dealing with changing dynamics. For example, multiple linear controllers that are switched or scheduled over the operating range are already used for fixed-wing aircraft [81], and other advanced techniques that are effective for multi-variable systems are being developed [20]. Several studies demonstrating the potential of these techniques for full-scale rotorcraft have also been presented [46, 40]. But the lack of accurate models has prevented their implementation in RUAVs.

3.2 Rotorcraft Control

To understand the main issues in controlling rotorcraft, we first briefly explain the role of the control inputs and the physical principles governing rotorcraft control.

Basic Rotorcraft Control Principles

The primary function of the four principal rotorcraft commands are the following: the main rotor lateral and longitudinal cyclic inputs control the roll and pitch moments produced by the main rotor; the collective input controls the magnitude of the main rotor thrust; the tail rotor collective input (pedal) controls the tail rotor thrust. Hence, the commands have a direct effect on the rotorcraft roll and pitch attitude rate, vertical velocity, and heading rate, respectively.

The pilot does not control the helicopter position or velocity directly, but via a chain of effects that can be summarized as follows. The cyclic control inputs result in control moments about the rotor hub via a tilting motion of the rotor disc (the rotor disc is used as a simplified representation of the combined effect

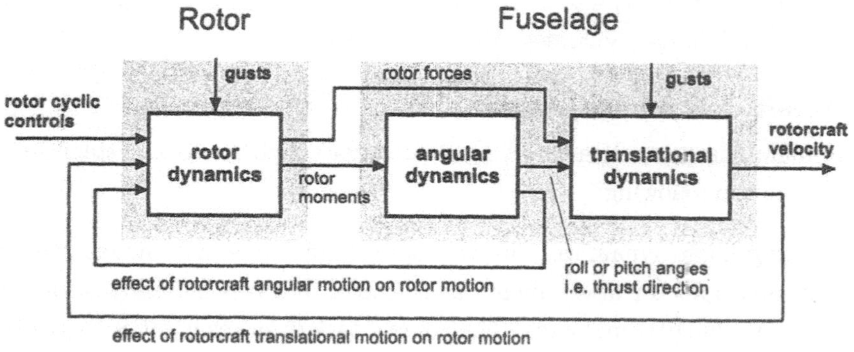

Figure 1.4. **Block diagram of the subsystems involved in the longitudinal-lateral helicopter control problem. Notice that the pilot does not have direct control over the vehicle translational motion, but instead via a chain of events involving the rotor and fuselage angular dynamics. At the same time, the pilot must compensate the secondary responses to his control action, the influence of translational velocity on the attitude, as well as effects of gust disturbances.**

of individual blade motion). The rotor control moments produce a fuselage rolling or pitching motion. If the helicopter is hovering, changing the fuselage's roll or pitch angle will result in a tilting of the rotor thrust vector, producing horizontal thrust components that acts as a propulsive forces. For example, by holding a constant pitch angle, the helicopter will accelerate until the propulsive force is balanced by the aerodynamic drag force. In steady state, a pitch or roll angle translates into a steady longitudinal or lateral velocity. Figure 1.4 shows a block diagram of the subsystems involved in the longitudinal-lateral helicopter control problem.

Of course, other effects will also need to be compensated, such as the effect of the translational or rotational fuselage motion on the rotor and fuselage attitude dynamics, or the direct secondary effects of the controls. Some of these secondary, or cross-axis, effects are easy to understand. For example, in the longitudinal velocity control, when the helicopter is pitched, the vertical thrust component will decrease, requiring an increase in the thrust magnitude to keep the vehicle at level altitude. This increase in thrust, however, will produce a reaction torque at the rotor shaft that in turn will result in a yawing moment, for which the pilot will need to adjust the tail rotor thrust. Other effects are more subtle, such as the roll responses following longitudinal cyclic and collective control actions and the pitch responses following lateral cyclic control actions.

Finally, the helicopter is also sensitive to atmospheric disturbances that affect the rotor and fuselage dynamics.

Rotorcraft Control Challenges

The main characteristics that render rotorcraft flight control system design difficult are the following:

- Changing flight dynamics: The helicopter flight dynamics change over the flight envelope. In hover flight, a helicopter behaves similarly about the lateral and longitudinal axes. The control task in hover flight is to position the helicopter in the Cartesian coordinate system. The four degrees of freedom (DOF) are the longitudinal, lateral, vertical positions, and heading. The four inputs give independent control over these four DOF. In contrast, in forward flight, the helicopter's behavior is closer to that of a fixed-wing aircraft: the helicopter is oriented according to the direction of motion; the longitudinal cyclic control input acts like an elevator; the lateral cyclic input like ailerons; the pedals like a rudder; and the collective control allows altitude changes independant of pitch angle. The control task in cruise flight is to track a trim trajectory, such as rectilinear or steady turn, or maneuver from one trim to another. Similar to fixed-wing aircraft, the vehicle's behavior can often be divided into lateral-directional and longitudinal-vertical dynamics. Because of the changes in the dynamic characteristics, a linear hover control system will not be effective beyond a certain airspeed, or vice-versa.

- Multivariable behavior: The helicopter, with its six degrees of freedom and four control variables, is a multivariable system that exhibits coupling effects among the different axes of motion. Precise coordination of the control inputs is necessary to achieve a definite helicopter response.

- Instability: The helicopter attitude dynamics are conditionally stable; a minimum amount of attitude feedback is required for the system to be stable; however, too much feedback will destabilize the system.

- Higher-order dynamics: Helicopters, unlike most fixed-wing aircraft, cannot be approximated accurately by rigid-body dynamics. The rotor, which can be regarded as the main source of forces and moments acting on the fuselage, responds both to pilot control inputs and fuselage motions. The resulting dynamic coupling between the rotor and the fuselage motion is a challenge for high-bandwidth control. Other higher-order dynamics that are sometimes important include higher-order rotor flapping, the lead-lag rotor degrees of freedom, the engine drivetrain system, the rotor inflow dynamics, and plant augmentations such as a stabilizer bar.

- Disturbance sensitivity: The main rotor is sensitive to atmospheric disturbances. Wind gusts will change the aerodynamic conditions at the rotor, and thus affect the local blade lift. Because the effect of wind gusts is similar to a local change in blade pitch angle, gusts act as input disturbances. Input disturbances are challenging because they are detected only after they have affected the vehicle response. Unmeasurable input disturbances require high control bandwidth to be effectively rejected.

From a practical standpoint, it is desirable to have a high level of control performance in a large portion of the helicopter's flight envelope and to be able to operate in windy or disturbed conditions, such as flying close to structures. These requirements can only be adequately addressed with multivariable design methodologies that can explicitly account for high-bandwidth dynamics. For optimal performance throughout the flight envelope, multiple design points with a gain scheduling/switching scheme is necessary.

Current Rotorcraft Flight Control Research

Small-scale rotorcraft have the same basic configuration and control modality as their full-scale counterparts. However, the control-design problem for small-scale unmanned rotorcraft has significantly different objectives and challenges than those occupying the traditional manned rotorcraft control field. Traditional rotorcraft control has been mostly concerned with improving the handling qualities of the aircraft for piloted operation; the control systems are primarily stability-augmentation systems (SAS) and low-authority autopilot systems, such as altitude- or attitude-hold systems. Helicopters present many interesting control challenges which make them excellent candidates for modern multivariable design methodologies. Some research using robust control methods exist, for example there is the line of work by Postlethwaite with H_∞ control [96, 90, 78], however, SAS systems are mostly based on classical feedback control.

Fully autonomous flight necessitates high-authority control systems, such as position-control systems and trajectory-tracking systems. Compared to the former area, little research has been conducted on these autonomous control problems. Autonomy is not unique to small-scale vehicles. Since the late 1990s, some autonomous rotorcraft projects utilizing full-scale rotorcraft have been developed for various military tasks. For example, an unmanned Kaman K-Max helicopter has been equipped for automated cargo delivery [43, 13], and a Schweizer 300 helicopter has been modified by Northrop Grumman and Ryan for airborne surveillance and communication relay for Navy and Marine operations [14]. These projects are often based on the control methodologies used for manned vehicles.

Some small-scale vehicles exhibit dynamic capabilities (high angular rates and high thrust-to-weight ratio) that far exceed those of full-scale vehicles [58], permitting non-traditional type of flight operations, such as high-incidence flight and even inverted fight. It is mainly the problem of autonomous control and high maneuverability that define today's unmanned rotorcraft research field. Work in this area has in large part developed in academia independent of the rotorcraft community.

For example, in Weilenmann's work [92, 94, 93] a rigged model helicopter was used as a test bed to evaluate the performance of various multivariable control design techniques (LQ, H_∞, H_2, and μ-synthesis). A classical single-input single-output (SISO) proportional-integral-derivative (PID) controller was used as a benchmark. The controller performance was compared using performance and stability metrics (e.g. closed-loop bandwidth, cross-axis effects, disturbance rejection, stability margins). The results show that the multivariable model-based control-design methods outperform the classical SISO control systems in controlling a helicopter.

The state of autonomous rotorcraft control at the end of the 1990's had not evolved much from the early work. In spite of the variety of control design methodologies tried, from classic feedback to robust, nonlinear [48], fuzzy [21], and adaptive techniques [72], no progress in the flying capabilities were achieved. The few flying vehicles were still capable of only basic autonomous control such as holding a hover flight and tracking modest position changes [3, 73, 75]. It is only recently, after the results of the present research were made available that significant results have been demonstrated in flight. Chapter 7 gives an overview of the most recent results.

Like in other control applications, there is always the question of which control design methodology will deliver the performance that satisfies the requirements for autonomous flight. It is fair to say that in the area of autonomous rotorcraft, more research has been performed on control design methodologies than in modeling. This tendency of trying to circumvent the modeling problem by searching for the perfect control methodology is not productive and can even lead to misleading conclusions about the effectiveness and applicability of certain approaches or techniques.

4. Material Preview

Now that we have a good motivation for using small-scale rotorcraft, have a sense of the challenges and difficulties in controlling them, reviewed the state of the art in rotocraft modeling and control, we better understand what is required to advance small-scale rotorcraft control design, and what will help set foundations for future research. In the following, I will give a short statement of objectives and then give a preview of the material of this book.

4.1 Statement of Objectives

The primary goal of this work is the development of a linear state-space model that accurately captures the key dynamics of small-scale rotorcraft with a minimum level of complexity. We want a model that we can use with a variety model-based control design methods. Moreover, we are interested in a model which describes the key physical effects in a simple but meaningful way, so that it can easily be understood and used by other researchers working with different small-scale vehicles. The model should, in fact, allow the identification of other small-scale rotorcraft, using a variety of identification methods, and also allow the extraction of physical information from the identified model.

In the following we describe the modeling process that we followed to accomplish this goal and include a description of some of the key characteristics of small-scale rotorcraft including the effects of the stabilizer bar and the effects of scaling. Finally to illustrate and demonstrate the model's suitability for control design we proceed in the form of a case study where we analyze the performance of a classic proportional-derivative (PD) attitude control system. For the control analysis we again use system identification, this time on the closed-loop system. This extra application of system identification demonstrates its essential role in the full control design cycle: from modeling and analysis of the vehicle's dynamics to controller design, analysis and validation.

4.2 Book Outline

In the following we provide a short preview of the material presented in the book. This will allow a quick and broad view of the covered topics.

Chapter 2

Chapter 2 describes the working principles of frequency-domain identification focusing on the tools and procedure used in CIFER. It also provides some theoretical background to frequency-response estimation, which corresponds to the "nonparametric" step of the identification. A key task in any identification is the collection of the experimental input-output data. The quality of the data determines to a large extent the results that can be obtained using system identification. Data quality depends on the instrumentation but it also depends critically on the flight-test techniques used for the data collection. For this work, the same frequency sweep flight-test technique that is recommended for large-scale-rotorcraft identification were used. The second part of Chapter 2, describes the hover and cruise flight experiments and presents the frequency responses that were estimated from the flight data. The frequency responses fully describe the linear dynamics of a system giving us essential information about the nature of a vehicle's dynamics.

Chapter 3

The key task in system identification is the development of a parameterized model of the vehicle dynamics. This development is performed from first principles and insight that can be gathered from the estimated frequency responses. Basically, for the model to successfully capture the information contained in the estimated frequency responses, it must describe the dynamic effects that come into play in the frequency range over which the model is fitted. Both for control design and system identification, it is advantageous, provided it does not overly hurt the model accuracy, to have a model of the lowest possible order and with the fewest unknown parameters. To satisfy the accuracy requirement for high bandwidth rotorcraft control [85], the model must typically account for effects beyond the rigid-body dynamics. In full scale rotorcraft these include the coupling between the fuselage and the rotor, the inflow, and the engine-drivetrain dynamics. Since small-scale rotorcraft have the same basic arrangement as their full-scale counterparts, we developed our model following the example of full-scale rotorcraft. However, it was not known *a priori* which effects were critically relevant in small-scale rotorcraft. Small-scale rotorcraft are also often equipped with a stabilizer bar and an active yaw damping system to help the pilot control the faster dynamics of such small vehicles. Since both systems strongly influence the vehicle dynamics, we needed to take them into account in the parameterized model. The chapter reviews some basic elements of rotorcraft dynamics focusing on the simplified rotor model, the coupling between the rotor and fuselage, and describes the approach used to account for the stabilizer bar system.

Chapter 4

Chapter 4 presents the identification of the parameterized model. It explains the setup of the identification including the output equation used to fit the model-predicted and estimated frequency responses, and shows how to account for sensor kinematics. Since the entire model cannot easily be identified at once, the identification is first broken down into smaller problems involving subsystems. The identified subsystems are then combined incrementally. During this process, derivatives are added or dropped based on performance statistics (sensitivity and Cramer Rao Bound). This identification process is used to determine the final parameterization of the model.

The accuracy of the final model – its ability to predict the vehicle response – is then verified by comparing the model-predicted responses with the responses recorded during separate flight experiments. Since system identification is a data-fitting process, even an accurate model does not imply that the model is physically meaningful, i.e. that the values identified for the derivatives correspond to the values predicted by theory. Having a physical meaningful model is important if we want to use the identification process to extract information about the vehicle and then use this to improve our understanding of small-scale rotorcraft dynamics. I verify this consistency by comparing the identified values for the key model parameters with those predicted by rotorcraft theory.

Chapter 5

Next, in Chapter 5, we use the identified model to describe the key characteristics of the attitude dynamics. In this chapter I introduce identification results from MIT's X-Cell helicopter. This smaller helicopter is used here to illustrate the effect of scale and rotor head design. I then apply dimensional analysis for a more precise understanding of the effects that scaling a rotorcraft has on a range of flight characteristics. I specifically look at two scaling hypothesis: Froude scaling (or dynamics similarity), and Mach scaling (speed similarity). The scaling rules derived from the two scaling hypothesis are then used to relate the dynamic characteristics of the Yamaha R-50 with those of a full-scale UH-1H helicopter and those of the X-Cell. I also briefly look at the effects the stabilizer bar has on the vehicle dynamics and finally describe the modal characteristics of the R-50 helicopter in hover and cruise flight.

Chapter 6

An accurate model like the one developed here lends itself to a range of control-design techniques. In the following will not try to propose a particular approach. Instead, the goal is to introduce the control problem from a broad physical perspective, presenting elements of analysis, design, optimization, and validation. Chapter 6 is presented as a case study. I first describe the proportional-derivative (PD) control system originally used on Carnegie Mellon's RUAV, hereby illustrating the baseline architecture of a rotorcraft control system. Subsequently I proceeded with the analysis of the performance of the attitude control system using both our model and closed-loop identification. Here, the identification provides for a mean to validate the model of the closed-loop system which combines the identified vehicle model with the elements of the control system. Using gain and phase margin analysis we can show that the performance of such a controller is seriously limited for helicopters equipped with a stabilizer bar. I used the insight from the analysis to develop a dynamic compensation scheme (notch filter). Here again we used closed-loop system identification to validate the compensation.

To illustrate one possible use for the model, I use the CONDUIT control system optimization tool to design the compensated attitude control system. Finally, I introduce the notion of flying qualities and describe some of the metrics used to specify the handling qualities requirements of military rotorcraft. I use those to evaluate the control performance of the attitude control system, and discuss how such metrics are still meaningful for unmanned miniature rotorcraft.

Chapter 7

Since the completion of the bulk of this work, important results were achieved and demonstrated in controlling miniature rotorcraft. Chapter 7 first gives a final view of the material and results presented in this book and then gives an overview of these recent developments and results. At the same time, it provides an outlook on future challenges in the areas of modeling, control design and guidance of unmanned aerial vehicles, giving a more complete picture of the developing field of aerial robotics.

Chapter 2

FREQUENCY RESPONSE SYSTEM IDENTIFICATION

In the first part of this chapter, we introduce the basic principles of frequency-domain identification and briefly discuss the advantages that frequency-domain methods have over time-domain techniques for rotorcraft identification. Then we give a description of the CIFER identification tool used for this work. In the second part of the chapter, we describe the flight experiments that we performed for the data collection and present the frequency responses estimated from the flight data.

1. System Identification Modeling

The goal of system identification modeling is the determination of a mathematical description of a plant based on observations made on that plant. A variety of system identification techniques exist [54] and the characteristics of the particular modeling problem determines which is better suited.

System identification is essentially a data-fitting process. For example, in the classical output-error method an optimization algorithm adjusts model parameters to minimize the error between the responses collected from the vehicle and the responses predicted by the model. The fitting error can be defined in the time domain, in which case the model-predicted response is obtained by integrating the equations of motion for the same input sequence used during the collection of the experimental data. It can also be defined in the frequency domain, in which case the fitting error is simply the difference between the frequency responses estimated from the experimental data and the frequency responses calculated analytically from the model.

Flight Vehicle Identification

The main developments in flight-vehicle system identification happened during the 1970's with the introduction of the output-error method. An overview of flight-vehicle system identification is available in [31]. System identification is an important approach to modeling flight vehicles, in particular for deriving simple, but accurate, linear representations for control design and flying qualities studies. It very much represents a modeling approach in its own right, but can also be used in accord with detailed first-principle modeling. For example, identification is also used for the validation and refinement of complex simulation models.

For flight vehicles, the task of identification modeling is the determination of equations of motion which will accurately predict the vehicle responses. Often a good enough knowledge of relevant dynamics is available to derive equations of motion from first principles, however, for more complex vehicles, such models will contain sufficient uncertainty in the physical parameters and/or higher-order dynamics so as to do a poor job at predicting the vehicle response. System identification is then appropriate for determining these unknown or poorly known parameters as well as refine the model structure.

Rotorcraft System Identification

Rotorcraft are among the most challenging vehicles to identify. In the beginning, identification techniques were mostly based on time-domain methods, and were not very successful with rotorcraft [44] for reasons that will be explained later in this section. Effective rotorcraft system identification became possible with the use of frequency-domain techniques and became practical with the development of tools like CIFER (Comprehensive Identification from FrEquency Responses) [86]). An overview of helicopter system identification is available in [32]; and the historical background to CIFER is given in [84].

In the past decade a variety of rotorcraft programs have benefited from frequency domain system identification. One of the early applications was the tilt-rotor aircraft XV-15 [84] followed by numerous other examples, including the BO-105 [87], the UH-60 [23], and the SH-2G [89].

The work presented herein represents the first comprehensive application to a small-scale unmanned rotorcraft (Yamaha R-50 [60, 61], later also the Miniature Aircraft X-Cell .60 [27, 58]). Since then it has been applied to more exotic air vehicles such as Microcraft's 9-inch ring wing ducted fan [53].

Challenges of Rotorcraft Identification

Rotorcraft present a challenging application for system identification. The main difficulties are as follows:

- A rotorcraft is a system with multiple inputs and multiple outputs. The coupling effects generally preclude the separation of the dynamics into smaller, less complex, subsystems such as the longitudinal-vertical or lateral-directional ones that are often used in fixed-wing aircraft.

- The powerplant, rotors, gear boxes, and shafts produce high levels of vibrations that result in noisy acceleration and/or angular rate measurements. Filtering the noise from the measurements can result in the loss of important information.

- The vehicle's dynamic modes are markedly separated in their frequencies, such as the low-frequency rigid body dynamics and the high-frequency rotor dynamics. This is particularly an issue for time-domain identification where low-frequency information can outweighs higher-frequency information in the fitting error.

- The dynamic characteristics of rotorcraft change significantly from one operating condition to another. When using linear system identification, the data collection must be performed for specific operating conditions. Multiple operating points need to be identified to characterize the vehicle dynamics in the entire operating envelope. Direct nonlinear identification methods are not popular, however, the individual linear models are sometimes used to tune a nonlinear model.

- Unstable dynamics impose constraints on the length and type of inputs used during data collection. For longer records, stabilizing feedback actions from the pilot or control system are necessary to prevent the aircraft from going out of control. However, as we will see later, these actions bias the estimation of the frequency responses.

2. Linear Frequency-Domain Identification

In this section, we first provide an overview of the linear frequency domain identification process, where we explain how it breaks down into different subproblems. We also briefly explain the advantageous properties of the frequency domain method. Then, we give some theoretical background on the estimation of frequency responses from input-output data; we explain how to account for accuracy requirements; and finally, we outline the development or the parameterized model used for the identification, which will be the main topic of next chapter.

2.1 Overview

Frequency-domain system identification is an output-error method where the data fitting takes place in the frequency domain. The parameters of the linear model, representing the vehicle's equations of motion, are identified by minimizing a frequency-domain cost function.

Working Principle

The frequency domain identification process divides into two parts: first, the estimation of frequency responses from the experimental data (nonparametric identification); second, the identification of the model parameters (parametric identification). An important part of this process is the development of the parameterized identification model and the refinement of this model until a satisfactory result is achieved. The validation of the model is usually performed in the time domain.

Figure 2.1 illustrates the working principle of frequency domain identification. The model is fitted to the frequency responses estimated from the input-output data over a pre-specified frequency range. This range is chosen based on the modeling requirements (model bandwidth) and is constrained by the quality of the flight data, i.e. how accurately the responses were estimated along the various frequencies. The fitting frequencies also correspond to particular physical effects occuring in the vehicle dynamics. The fitting range defines what we could call the effective dynamics of the system. The model order and the model parameterization must be consistent with these effective dynamics. This defines a form of equivalence between the model complexity (model order and number of unknown parameters) and the amount of information available from the collected input-output data. When possible, it is best to derive the parameterized model based on an analysis of the estimated frequency responses. This will reduce the number of model refinement steps in the identification process.

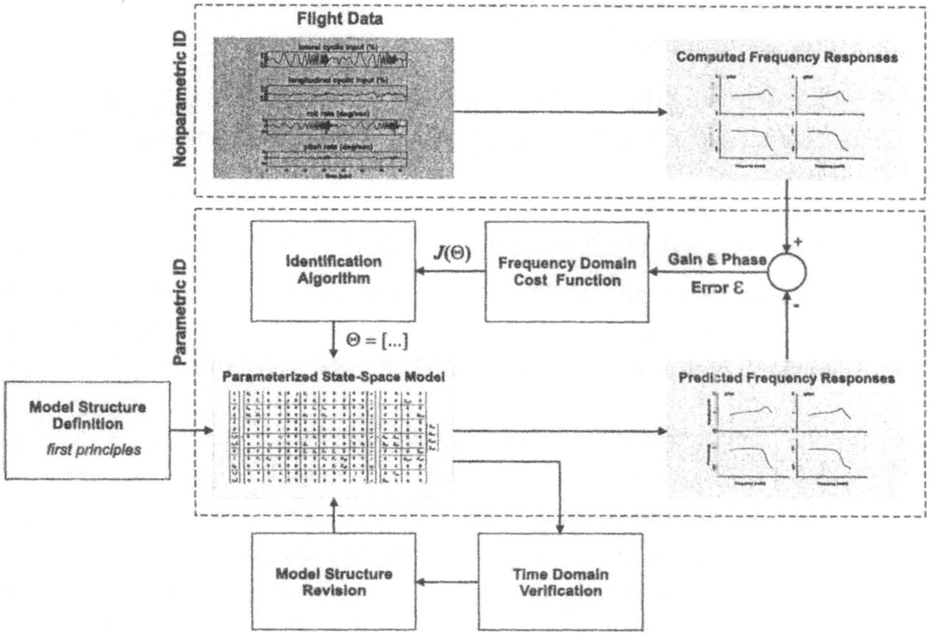

Figure 2.1. Flowchart of the frequency-domain identification process.

Advantages of Frequency Domain Methods

Frequency-domain identification has properties that make it better suited to rotorcraft than time-domain identification. The following is a list of key advantages:

- Output measurement noise or process noise that is uncorrelated write the control inputs does not bias the frequency-response estimates. Hence this type of disturbance is automatically separated from the data.

- It is possible to individually specify the frequency range used for the fitting of each input-output pair. This is particularly effective in addressing the frequency disparity between the dynamic modes, allowing improvements in the identification of the parameters belonging to both the quasi-steady rigid-body dynamics and those parameters belonging to the faster rotor dynamics.

- Selection of specific frequency ranges is also effective at separating information that is relevant for the identification of the vehicle dynamics from irrelevant information such as structural modes or even disturbances, which are often characterized by their frequency content.

- The formulation of the cost function in the frequency domain requires fewer data points than in the time domain, resulting in a more efficient parameter identification.

2.2 Theoretical Backround

In the following we provide basic theoretical foundations to frequency response function and their estimation from sampled input-output data. For details on these topics see the classic signal processing references such as [66].

Frequency Response as Nonparametric Model

Frequency responses fully describe the linear dynamics of a system (when the system has nonlinear dynamics, system identification determines the describing functions [77], which are the best linear fit of the system response).

In the time domain, the input $u(t)$ and output $y(t)$ of a linear system can be related to each other through the convolution integral [41]:

$$y(t) = \int_{-\infty}^{\infty} h(\tau)u(t-\tau)d\tau \tag{2.1}$$

where $h(\tau)$ is the impulse-response function (we assume here that the system is time invariant, i.e. $h(\tau, t) = h(\tau)$, and that it is physically realizable, i.e. $h(\tau) = 0$ for $\tau < 0$). In the frequency domain the input and output of a system are related through the frequency-response function $H(j\omega)$ [41, 5]:

$$Y(j\omega) = H(j\omega)U(j\omega) \tag{2.2}$$

where $Y(j\omega)$ and $U(j\omega)$ are the Fourier transforms of the output and input signals $y(t)$ and $u(t)$, respectively:

$$Y(j\omega) = \int_{-\infty}^{\infty} y(t)e^{-j\omega t}dt \tag{2.3}$$

$$U(j\omega) = \int_{-\infty}^{\infty} u(t)e^{-j\omega t}dt. \tag{2.4}$$

The frequency-response function $H(j\omega)$ is the Fourier transform of the impulse-response function $h(\tau)$:

$$H(j\omega) = \int_{0}^{\infty} h(\tau)e^{-j\omega t}d\tau. \tag{2.5}$$

$H(j\omega)$ is a complex quantity with a real $H_R(j\omega)$ and an imaginary part $H_I(j\omega)$

$$H(j\omega) = H_R(j\omega) - jH_I(j\omega). \tag{2.6}$$

The complex frequency-response function can be represented in polar notation:

$$H(j\omega) = |H(j\omega)|e^{-j\phi(j\omega)} \tag{2.7}$$

where $|H(j\omega)|$ is the magnitude and $\phi(j\omega)$ is the phase of the frequency response. These are defined by:

$$|H(j\omega)| = [H_R^2(j\omega) + H_I^2(j\omega)]^{1/2} \tag{2.8}$$

$$\phi(j\omega) = \tan^{-1}\left[\frac{H_I(j\omega)}{H_R(j\omega)}\right]. \tag{2.9}$$

The frequency response fully describes the dynamic characteristics of a linear time-invariant (LTI) system. For a multi-input multi-output (MIMO) system, H is a $(n_o \times n_c)$ matrix of frequency responses relating the n_c inputs to the n_o outputs.

Calculation of Frequency Responses Estimates

The complex frequency response function $H(j\omega)$ also relates the input auto-spectral density function G_{uu} to the input-output cross-spectral density function G_{uy}, i.e.

$$G_{uy}(j\omega) = H(j\omega)G_{uu}(j\omega) \tag{2.10}$$

where G_{uu} can be defined via the Fourier transforms:

$$G_{uu}(j\omega) = \lim_{T\to\infty}\frac{2}{T}E[|U_k(j\omega,T)|^2] \tag{2.11}$$

and G_{uy} is defined by:

$$G_{uy}(j\omega) = \lim_{T\to\infty}\frac{2}{T}E[U_k^*(j\omega,T)Y_k(j\omega,T)]. \tag{2.12}$$

From Eq. 2.10 we see that for a single-input single-output system (SISO), the frequency-response function can be determined from the spectral functions, i.e.

$$H(j\omega) = \frac{G_{uy}(j\omega)}{G_{uu}(j\omega)}. \tag{2.13}$$

For a MIMO system, we have a vector of n_c frequency responses relating the n_c inputs to the *j-th* ouput y_j

$$\boldsymbol{H}_j(j\omega) = \begin{bmatrix} H_{y_ju_1} \\ H_{y_ju_2} \\ \vdots \\ H_{y_ju_{nc}} \end{bmatrix} = \boldsymbol{G}_{uu}^{-1}\boldsymbol{G}_{uy_j}, \tag{2.14}$$

where $\boldsymbol{G}_{uu}$ is the $n_c \times n_c$ matrix of auto- and cross-spectral density functions of the n_c input signals and $\boldsymbol{G}_{uy_j}$ is the vector of cross-spectral density functions between the n_c inputs and the output y_j. Equation 2.14 represents the conditioned SISO spectral functions. The frequency responses of a MIMO system are only equivalent to the SISO frequency responses in the absence of coupling in the system and correlation among the system inputs [5].

The spectral auto- and cross-density functions can be derived either via the Fourier transform of the auto- and cross-correlation function ($R_{uu}(\tau)$ and $R_{uy}(\tau)$), via analog filtering, or via the Fourier transforms [5]. In the following, we will describe the third approach.

We assume that we have N_d input and output segments (i.e. collected from N_d experiments), each of length T_d seconds. The sampling interval is T_s; each segment therefore contains $N = T_d/T_s$ samples $u(t_n)$ and $y(t_n)$ ($n = 0, 1, 2, \ldots, N-1$).

The discrete equivalent for the Fourier transform of Eqs. 2.3 and 2.4, used for sampled data, is the discrete Fourier transform (DFT)

$$Y(j\omega_k) = \sum_{n=0}^{N-1} y(t_n) e^{-j\omega_k t_n}, \; k = 0, 1, 2, \ldots, N-1 \tag{2.15}$$

$$U(j\omega_k) = \sum_{n=0}^{N-1} u(t_n) e^{-j\omega_k t_n}, \; k = 0, 1, 2, \ldots, N-1 \tag{2.16}$$

where $\omega_k = k\Omega_s$ are the discrete frequency points (frequency samples). The frequency sample interval is given by

$$\Omega_s = \frac{2\pi}{NT_s}. \tag{2.17}$$

In CIFER, the DFT computation is performed using a Chirp-Z algorithm [87]. This algorithm offers some numerical advantages over other DFT such as the fast Fourier transform (FFT) (e.g. specify frequencies on nonuniform grid-points).

Estimates of the spectral density functions that are needed to compute the estimates of the frequency responses are computed from the N_d finite Fourier transforms [5]

$$\hat{G}_{uu}(j\omega_k) = \frac{2}{N_d T} \sum_{n_d=1}^{N_d} |U_{n_d}(j\omega_k)|^2 \tag{2.18}$$

$$\hat{G}_{uy}(j\omega_k) = \frac{2}{N_d T} \sum_{n_d=1}^{N_d} U_{n_d}^*(j\omega_k) Y_{n_d}(j\omega_k). \tag{2.19}$$

Finally, the estimates of the frequency responses at the discrete frequency points ω_k are computed from

$$\hat{H}(j\omega_k) = \frac{\hat{G}_{uy}(j\omega_k)}{\hat{G}_{uu}(j\omega_k)}. \tag{2.20}$$

For MIMO systems the discrete version of Eq. 2.14 is used.

Coherence Function

An important metric that can also be determined from the spectral density functions is the coherence function γ_{uy}. The coherence γ_{uy} (partial coherence for a MIMO system) indicates how much an output y is linearly correlated with a particular input u as a function of frequency. Following the simpler SISO case, the coherence is computed from the cross-spectrum G_{uy} and the input and output auto-spectra G_{uu} and G_{yy}, respectively

$$\gamma_{uy}^2 = \frac{|G_{uy}|^2}{G_{uu}G_{yy}} \leq 1. \tag{2.21}$$

Note that for the MIMO case, the partial coherence is derived from the conditioned spectrum. The coherence is essentially a correlation metric defined in the frequency domain. It is normalized so that when two signals are fully correlated at a frequency point, the coherence function will attain a maximum value of unity. Conversely, when completely uncorrelated, the function will be zero.

Errors in the Frequency Response Estimates

Frequency-response estimates are subject to random errors and bias errors. Random errors are due to measurement noise and unmeasured input disturbances that are uncorrelated with the inputs (e.g. wind gusts). Bias errors are due to noise in the input measurement (which does not pass through the system), nonlinear system characteristics or unmeasured inputs that are correlated with the input.

There is a relation between the coherence function $\gamma(\omega)$, the random errors in the frequency-response estimates, and the number of averages performed in the calculation of the spectral density estimates (see [5]). The normalized error in the frequency-response magnitude is given by:

$$\varepsilon[|\hat{H}(\omega)|] \approx \frac{[1-\gamma^2(\omega)]^{1/2}}{|\gamma(\omega)|\sqrt{2N_d}}, \tag{2.22}$$

where N_d is the number of sequences (i.e averages) used in the estimation of the spectral functions. The error σ in the phase (in radians) is given by:

$$\sigma[|\hat{\phi}(\omega)|] \approx \sin^{-1}\left(\varepsilon\left[|\hat{H}(\omega)|\right]\right). \tag{2.23}$$

From Eq. 2.22 we see that averaging is important to reducing the estimation error, i.e. in particular more averages should be used if the quality of the experiments and measurements is poor. For example, by using five time histories for the estimation of the frequency response ($N_d = 5$), a coherence $\gamma^2 \leq 0.6$ would give a normalized magnitude error of $\varepsilon[|\hat{H}(\omega)|] \approx 0.26$ (26%) and a phase error of $\sigma[|\hat{\phi}(\omega)|] \approx 0.26$ rad (14 degrees). In comparison, a single time history ($N_d = 1$) with the same coherence would result in 58% magnitude and 35 degrees phase error.

In CIFER, the coherence is used as a weighting function in the frequency-response fit (output-error cost function Eq. 2.26) during identification of the model parameters. The user can also specify a cutoff value for the coherence to discard the frequency points in the cost function which have coherence less than the cutoff value.

2.3 Model Accuracy Requirements

For control design applications, we are concerned about the frequency range over which the model is accurate. For feedback control systems, the model used for the design should be accurate at frequencies well beyond the target crossover frequency ω_c. For example, according to [85], dynamic modes within 0.3–3 times the target crossover frequency contribute substantially to the characteristics of the closed-loop system.

There is a precise relation between the model accuracy level (model uncertainty) and the control performance that can be attained with a controller. The performance vs. robustness tradeoff is the central topic of robust control theory (see [18]). It is interesting to note that the model's magnitude and phase errors given by Eqs. 2.22 and 2.23, and the fitting error achieved by the identified parameterized model could be used explicitly in the analysis and design of a robust control system. System identification, besides suporting the modeling task, also provide a quantitive measure of the accuracy of the model.

2.4 Development of a Parameterized Model

An important part of the identification problem is the development of the parameterized model of the vehicle dynamics. This model is usually developed from first-principles. The vehicle equations of motion are then linearized and simplified. Since the ultimate purpose of the model is to enable the best possible fit of the estimated frequency responses, the responses should be used to guide the development of the model. For rotorcraft, the main question is usually how many effects beyond the rigid-body dynamics need to be accounted for.

The developement of the model continues even during the fitting process. Thus, modeling using system identification should be regarded as an incremental process, where the final model parameterization and structure is only established once the identification has been successfully completed.

For our application, the linear parameterized model is in the form of a state-space equation

$$\dot{\boldsymbol{x}}(t) = \boldsymbol{A}(\Theta)\boldsymbol{x}(t) + \boldsymbol{B}(\Theta)\boldsymbol{u}(t - \tau_i) \tag{2.24}$$

$$\boldsymbol{y}(t) = \boldsymbol{C}(\Theta)\boldsymbol{x}(t) + \boldsymbol{D}(\Theta)\boldsymbol{u}(t - \tau_i). \tag{2.25}$$

where the system matrices $\boldsymbol{A}$ and the input matrix $\boldsymbol{B}$ are functions of the unknown parameters (contained in the vector Θ). In addition to the parameters of the state-space matrices, individual time delays τ_i associated with each system input u_i can be included as part of the unknown parameters.

2.5 Parameter Identification

The unknown model parameters are identified through an optimization process in which the frequency-response fitting error is minimized. For this process, a cost function that measures how well the frequency responses calculated from the parameterized model fit the estimated frequency is built. The cost function depends on the unknown parameters Θ and is calculated as the sum of the magnitude and phase errors evaluated over n_ω discrete frequency points $\omega_1, \omega_2, \ldots, \omega_{n_\omega}$ (default number in CIFER is $n_\omega = 50$). It has the general form

$$J(\Theta) = \sum_{i=1}^{n_\omega} \boldsymbol{\varepsilon}(\omega_i, \Theta)^T W(\omega_i) \boldsymbol{\varepsilon}(\omega_i, \Theta) \tag{2.26}$$

where $\boldsymbol{\varepsilon}(\omega_i, \Theta)$ is the vector of magnitude and phase errors. The importance of the magnitude relative to the phase error can be specified (The default value is 20dB for 57.3 degrees verify). $W(\omega_i)$ is the frequency-dependent weighting function that emphasizes the frequency points at which the response is most accurate. Note that $W(\omega_i)$ is usually determined by the coherence function $\gamma^2(\omega_i)$.

The model-predicted frequency responses are determined from the model's transfer functions $T_{uy}(s)$. Applying the Laplace transform to the state-space equations and solving for the output vector $\boldsymbol{y}(s)$ yields the following expression for the transfer function:

$$\boldsymbol{T}_{uy}(s) = \frac{\boldsymbol{y}(s)}{\boldsymbol{u}(s)} = \left[\boldsymbol{C}[sI - \boldsymbol{A}]^{-1}\boldsymbol{B} + \boldsymbol{D}\right] (e^{-s\tau_i}) \tag{2.27}$$

For a MIMO system, $\boldsymbol{T}_{uy}(s)$ is a matrix of transfer functions, where each element relates a particular output y_j to a particular input u_i.

The parameters Θ are identified through the minimization of the cost function $J(\Theta)$. This process represents a nonlinear optimization problem that is usually solved using a secant search algorithm [54].

3. CIFER System Identification Tool

CIFER comprises a set of tools to perform frequency-domain identification [87]. The tools were originally developed for rotorcraft identification but can also be used for other systems for which a frequency-response approach is indicated, such as for the modeling of structural dynamics. All the tools are integrated around a database that conveniently organizes the large quantity of data generated throughout the identification process. The main tools included in the CIFER system are:

- *Frequency-response estimation (FRESPID).* Computes the frequency response for each input-output pair using a special FFT algorithm (Chirp-Z). Different window length are used in this process. At the same time, it computes the associated coherence functions. These frequency responses are then stored in the database to be processed by the following tools.

- *Multivariable frequency-response conditioning (MISOSA).* Conditions the single-input single-output frequency responses from FRESPID. This process removes the effects from the secondary inputs. At the same time it computes the "partial" coherences, which only reflect the effect of the primary input.

- *Optimal window combination (COMPOSIT).* Combines frequency responses computed in FRESPID using different window length to improve the accuracy of the low, middle, and high frequency ends. This process is solved through an optimization which finds the weighted sum of the individual frequency response that maximizes the coherence across the entire frequency range.

- *Transfer function identification (NAVFIT).* This utility allows one to identify the parameters of a transfer function using a specified frequency response. The user can specify the order of the numerator and denominator, and whether a time delay should be included.

- *State-space identification (DERIVID).* Identifies the parameters of a user specified state-space model using the selected frequency responses.

- *Time-domain verification (VERIFY).* Performs a time-domain verification of the model. The time responses predicted by the identified model are compared with the actual vehicle responses collected from flight experiments. Usually responses that are different from the ones used for the estimation of the frequency responses are used.

4. Flight Experiments and Data Collection

High quality flight data is essential to successful identification. The principal concerns are: the accuracy of the measurements (i.e. unbiased, disturbance free, no drop outs); the information content of the flight data; and the compatibility of the flight data with the premise of linear system identification. While the measurement accuracy depends on the instrumentation, the information content and compatibility depends mostly on the flight experiments used for the data collection.

4.1 General Flight-Testing Rules

The flight-testing rules are derived from theoretical and practical considerations. From the properties of coupled MIMO systems with correlated inputs it is possible to determine the experimental conditions that are most favorable to good frequency-response estimation (detailed analysis of SISO and MIMO frequency-response estimation is available in [5]). These conditions cannot typically be implemented entirely; a proper tradeoff must be found. For the small-scale rotorcraft, we followed the guidelines available for full-scale rotorcraft [95]. Later in this section, we will describe the actual flight test conducted on the R-50. The principal considerations are:

Sufficient excitation throughout frequency range of interest Dynamic modes within the frequency range of interest should be excited during the data collection. This is achieved by using frequency-sweep inputs (sinusoidal input with monotonically increasing frequency). Other types of broadband inputs exist (e.g. pseudo-binary-random or multi-step), however, they are not well suited to piloted experiments. Figure 2.2 shows a sample of two concatenated records collected during a lateral cyclic experiment. Both the recorded variables corresponding to the primary axis (lateral) and the secondary axis (longitudinal) are shown. Low-frequency excitations (0.1-2 rad/sec) are important for the identification of the quasi-steady effects, such as the aerodynamic derivatives. High-frequency excitations (8-14 rad/sec) are important for the identification of effects beyond the rigid-body dynamics such as the coupled rotor-fuselage dynamics and the stabilizer bar dynamics, as well as time delays.

Record length and sampling rate The length of each recorded data segment T_d determines the lowest-frequency component that can theoretically be estimated from the data, and the sampling interval T_s determines the upper limit [5]. The lowest frequency is given by $\omega_{\min} = 2\pi/T_d$, i.e. an entire period of the signal must be present in the data segment; for practical applications, however, a record length that is at least twice the period of the lowest desired frequency ($T_d = 4\pi/\omega_{\min}$) [84] is recommended. The highest frequency is prescribed by the Nyquist cutoff frequency $\omega_{max} = \pi/T_s$. Obviously the quality of the sensors and the way the sampling process is implemented (e.g. use of an anti-aliasing filter) are criticaly important.

Maintain the operation within range of linear dynamics By adjusting the magnitude of the excitation, the pilot ensures that the magnitude of the vehicle's response remains within a region where its dynamics are mostly linear, as well as within the operating conditions that are examined (e.g., hover, slow cruise flight, medium cruise flight). Small-scale rotorcraft, because of their higher maneuverability compared with their full-scale counterparts, should tolerate stronger control inputs. In general, the pilot can detect these limits and thus has a good sense of the acceptable range. Also, it is important to choose calm weather conditions, because wind gusts act as unmeasured inputs that can significantly degrade the quality of the frequency-response estimates.

Minimize correlation among inputs Correlation among control inputs biases the frequency-response estimates [5]. A possible source of correlation is pilot decoupling compensation, e.g. simultaneous compensation of the sweep input on the secondary axes. To minimize this effect, it is necessary to let the helicopter respond freely to the sweep excitation as much as possible. The low-frequency range of the test, where the helicopter can easily undergo large speed deviations, is the most challenging. In this situation, it is better to reduce the amplitude of the excitation than to try to compensate for its effect.

Minimize amount of feedback Feedback can mask the dynamics of the open-loop vehicle. This can be illustrated through the following simple example. Consider a SISO system with plant transfer functions $P(s)$ and a feedback control system with transfer function $C(s)$. By definition, the gain of the control system is large at frequencies below the crossover frequency ($|C(j\omega)| \gg 1$, $\omega < \omega_c$) and small at frequencies larger than ω_c ($|C(j\omega)| \ll 1$, $\omega > \omega_c$). Hence, the following can be said about the closed-loop transfer function in the two extremes of the frequency range:

$$G_{clp}(j\omega) = \frac{PC}{1+PC} \approx \begin{cases} 1/C & \text{for } \omega < \omega_c \\ P & \text{for } \omega > \omega_c \end{cases} \tag{2.28}$$

Hence, for frequencies well below the crossover frequency, the closed-loop dynamics are fully determined by the control system, i.e. the dynamics of the plant disappear. For frequencies well above the crossover frequency, the closed-loop dynamics are fully determined by the open-loop plant dynamics. Based on these considerations, if feedback compensation is absolutely necessary, the smallest posssible feedback gains should be used (minimum crossover frequency).

The quality of the flight data, in particular its suitability for linear system identification, can be determined from the coherence function γ. Looking at the estimated frequency responses together with the coherence can help identify problems with the experiments and/or the instrumentation, namely a low coherence can result from insufficient excitation (low signal-to-noise ratio), non-linear effects in the dynamics, or noisy measurements. Of course, it can also mean that this particular output is not influenced by that input; in this case, the magnitude of the frequency response should be small.

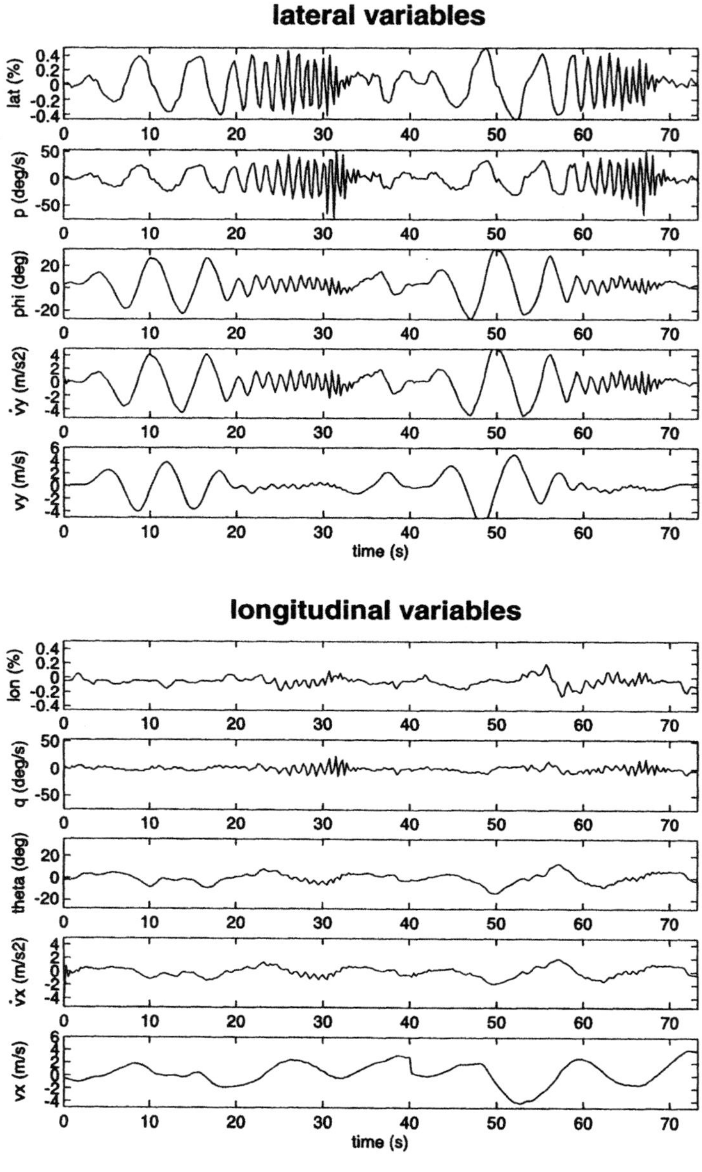

Figure 2.2. Sample of two concatenated records collected during a lateral cyclic experiment. Both the recorded variables corresponding to the primary axis (lateral) and the secondary axis (longitudinal) are shown. The top plot shows the lateral control input on which the frequency sweep was applied (note the increasing frequency of the input).

4.2 Description of the R-50 Flight Test

Two series of flight experiments were organized for each hover-flight and cruise-flight operating point. Note that it is important to train the pilot in the frequency sweep technique. For each flight, the pilot applied a frequency sweep sequence to one of the four control inputs via the remote control (RC) unit. While doing so, he used the other control inputs to maintain the helicopter close to the operating point of interest. In order to reduce the random error (Eq. 2.22), each experiment was repeated until four to five runs of good quality were collected.

The experiments were conducted open-loop, except for an active yaw damping system and the stabilizer bar, which can, as we shall see later, be regarded as a feedback system. The Yamaha R-50 also used an input cross-feed between the collective input and the pedal input to reduce the effect of changing main rotor torque on the helicopter yaw.

The control inputs used for the experiments were the stick deflection from the cyclic longitudinal δ_{lon} and cyclic lateral δ_{lat}, collective δ_{col} and pedal δ_{ped}. The vehicle variables (see Fig. 3.1) that were recorded are:

- Euler angles: roll ϕ, pitch θ, yaw ψ
- angular body rates: roll p, pitch q, yaw r
- body accelerations: a_x, a_y, a_z
- body velocities: u, v, w

During the experiment, all control inputs and all vehicle state variables were recorded and sampled at 100 Hz ($T_s = 0.01$ sec, i.e. $\omega_{max} \approx 31$ rad/sec). The raw flight data was then manually segmented to isolate each experimental run. Only the best segments were selected (the particular experiment corresponding to each segment was judged during flight, and the time histories of the vehicle states were subsequently also inspected). The data was filtered (-3dB @10 Hz) to remove effects of structural vibrations. The lengths of the data segments collected for the hover flight condition were all above 20 seconds ($T_d = 20$ sec, i.e. $\omega_{min} \approx 0.31$ rad/sec). For the cruise condition only 10 seconds long segments were feasible ($T_d = 40$ sec, i.e. $\omega_{min} \approx 0.63$ rad/sec).

Flight-Test Implementation

During the hover-flight experiments, the helicopter was near the pilot, and it was relatively easy to keep the helicopter within the hover operating region.

The cruise-flight experiments, however, were more problematic. We used a fly-over technique: the pilot accelerated the R-50 to a constant velocity of about 15-20 m/sec. The acceleration phase was performed over a specified distance, and was started at a location that would bring the vehicle directly above the pilot when the cruise speed was reached. At that point, the pilot held the velocity and started performing the different piloted sweeps while a rectilinear path above the airstrip.

We encountered two problems during the cruise-flight experiments. First, it was difficult for the pilot to maintain a precise airspeed, because he relied only on visual sighting for helicopter attitude information. Second, the length of the flight experiment was limited, because the helicopter had to be kept in sight. Attempts to increase the record length degraded the stability of the airspeed. By allowing the airspeeds to vary between 10 and 20 m/sec, we were able to achieve a maximum record length of 10 sec. This issue could be addressed in future experiments by conducting the flight testing on a paved runway and using an automobile as a chase vehicle, or even by performing closed-loop computer-controlled flight testing.

In general, collecting data that provides low-frequency information has been the biggest challenge. Due to the unstable vehicle dynamics, the pilot was forced to provide some minimum level of feedback in order to prevent the helicopter from going out of control. As a result, even long records (length above 40 sec used in hover) did not provide significant additional low-frequency information.

5. Analysis of the Estimated Frequency Responses

From the R-50 flight data, we estimated the frequency responses for all the vehicle's input-output pairs for both the hover- and cruise-flight condition. In the following, only the hover data is shown (the cruise frequency responses are shown in Chapter 4). Also, results for the X-Cell will be used mostly to illustrate the effects of scale presented in Chapter 5.

The units are a percentage of the full deflection available to the pilot for the inputs (all inputs: ±1 being $\pm100\%$ except the collective where the range is 0 to 1), rad/sec for the angular velocity and m/sec^2 for the acceleration. The primary responses to the cyclic control inputs δ_{lon} and δ_{lat} are the pitch and roll angular velocities q and p; the roll and pitch attitude angles ϕ and θ translate into lateral and longitudinal accelerations a_y and a_x. The primary response to the collective input δ_{col} is the vertical acceleration a_z, and, finally, the primary

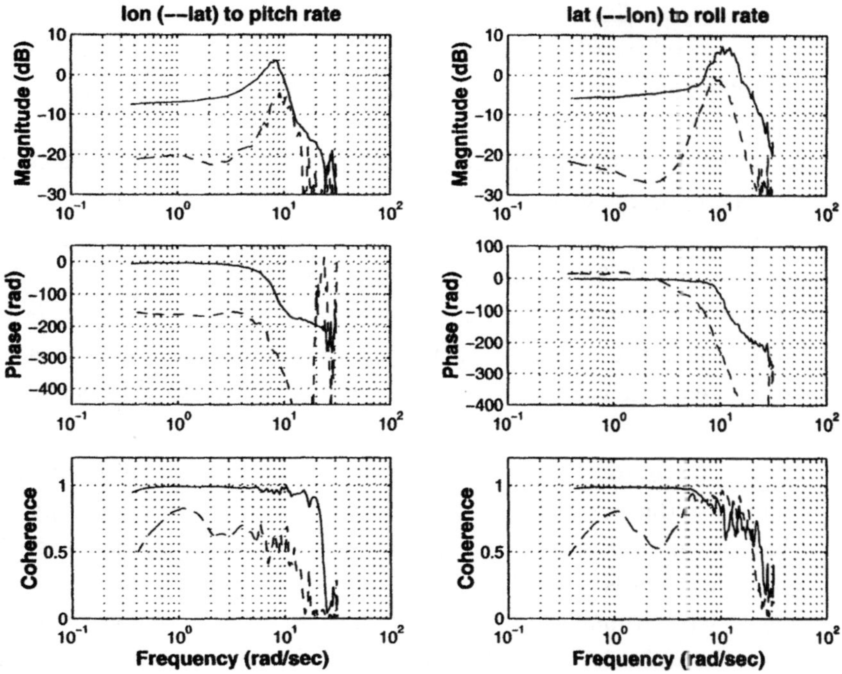

Figure 2.3. Estimated frequency responses for the pitch and roll rate response to longitudinal and lateral cyclic control inputs (CMU's Yamaha R-50)

response to the pedal input δ_{ped} is the yaw angular rate r. The analysis of the estimated frequency responses for these primary responses as well as the off-axes responses, provide insight into the flight dynamic characteristics of the aircraft. This insight will help us in the development of a parameterized identification model. The estimated frequency responses for the hover condition cover a frequency range from 0.32 to 31 rad/sec.

Roll and Pitch Responses

Figure 2.3 shows the estimated frequency responses for the on- and off-axes roll and pitch angular rates to the lateral and longitudinal cyclic inputs, as well as the values of the respective coherence functions.

Both on-axis pitch and roll rate responses q/δ_{lon} and p/δ_{lat} exhibit well-defined characteristics of a second-order system (magnitude roll off at 40 dB/dek and phase roll off of –180 deg). The magnitude increases slowly from the low frequency and peaks at around 8 rad/sec for the pitch response and 12 rad/sec for the roll response. The magnitude roll-off at low frequencies is due to the so-called speed stability effect, i.e. the long term attitude response is "washed out" by the effects of airspeed on the vehicle attitude. If lower-frequency data were available, this effect would be visible in a more dramatic way. The peak overshoot is about 10 dB (lightly damped mode). The resonant mode results from the dynamic coupling of the vehicle angular motion and the rotor flapping motion. The light damping of the fuselage-rotor mode is due to the presence of a stabilizer bar [35]. The coherence reaches unity almost up to 10 rad/sec. The coherence falls sharply due to the limited pilot bandwidth. In the lateral response, the coherence decreases earlier and more progressively, because the pilot needs to be cautious about overexciting the lightly damped fuselage-rotor mode.

The off-axis pitch and roll rate response q/δ_{lat} and p/δ_{lon} both have mid-range magnitudes that are on average 20 dB below, (i.e. one order of magnitude below) that of the on-axes responses. The off-axis responses have a similar second-order characteristic as the on-axes responses except for a drop in the mid-frequencies (at around 1 rad/sec). The coherence in the off-axes responses is on average between 0.5 and 0.7. The lower coherence is partly due to the small magnitude of the off-axis responses and hence low signal-to-noise values.

Body Longitudinal and Lateral Accelerations

Figure 2.4 shows the estimated frequency responses for the on- and off-axes longitudinal and lateral body accelerations to the lateral and longitudinal cyclic inputs, respectively, as well as the values of the respective coherence functions.

The on-axis acceleration responses a_x/δ_{lon} and a_y/δ_{lat} have the approximate shape of the integrated pitch and roll angular rate responses, respectively. This is physically meaningful, since the longitudinal and lateral forces components are proportional to the vehicle pitch and roll angle. Note that this relation is also confirmed by the 90-degree phase shift relative to the rate responses.

The pronounced resonance peak, which is mostly visible in lateral acceleration response, is in part also due to kinematic effects arising from an offset between the vehicle's center of gravity and the location of the inertial measurement unit. Because of this offset, a roll or pitch motion will produce tangential and centrifugal acceleration components at the sensor. This effect will need to be taken into account during the identification of the parameterized model.

The value of the coherence function for the on-axes responses is close to unity up to 3 rad/sec and drops progressively to reach zero at around 10 rad/sec. In

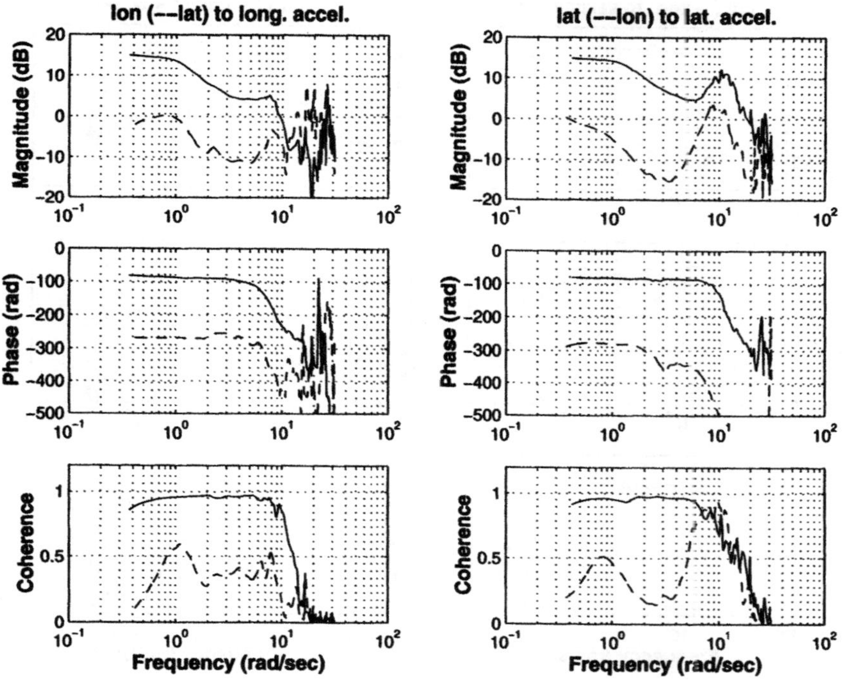

Figure 2.4. Estimated frequency responses for the longitudinal and lateral body accelerations to the longitudinal and lateral cyclic control inputs

the off-axes responses, the coherence has values below the 0.6 threshold except for the lateral acceleration (which could be due to the sensor position offset).

Yaw Rate and Vertical Body Acceleration

Figure 2.5 shows the estimated frequency responses for the yaw angular rate and vertical body acceleration to the pedal and collective inputs, respectively, as well as the values of their coherence functions.

The yaw rate response to the pedal input r/δ_{ped} exhibits the characteristics of a second-order system with a small resonance peak. These characteristics are due to the active yaw rate system (yaw rate feedback). The magnitude decreases towards the low frequencies due to the tail and tail rotor damping forces. The coherence function has a value close to unity through 10 rad/sec, with a dip at around 4 rad/sec, which coincides with the natural frequency of the second-order system.

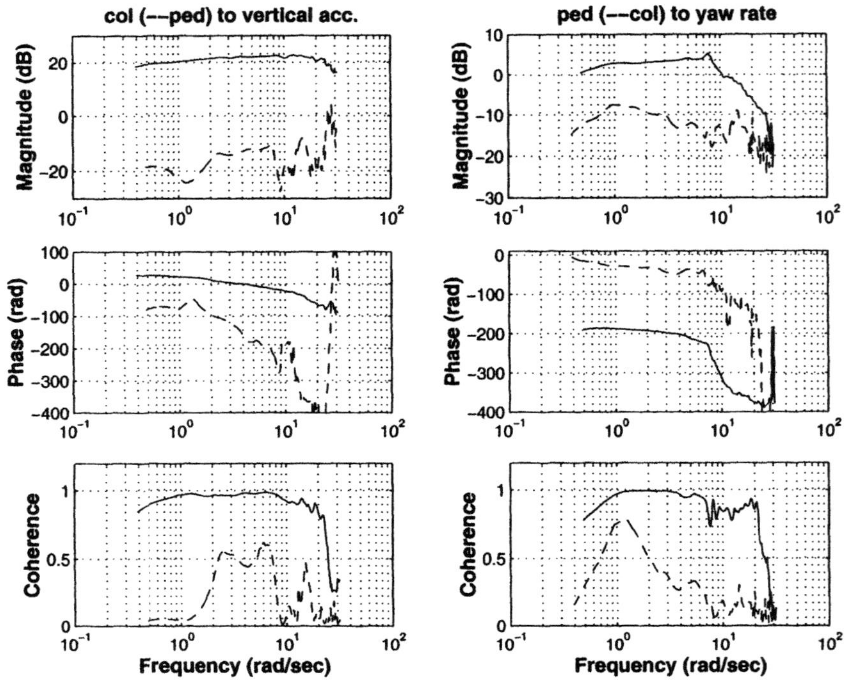

Figure 2.5. Estimated frequency responses for yaw rate response to pedal input and vertical body acceleration to collective control input

The vertical body acceleration response to collective input a_z/δ_{col} exhibits a flat response from the lowest estimated frequency up to about 10-15 rad/sec. The magnitude also shows a decreasing trend towards the low frequencies. The decreasing magnitude is again due to damping forces (here mostly rotor damping due to increasing inflow angle with increasing vertical speed). Note that this response does not exhibit effects of inflow dynamics or rotor coning dynamics usually visible in full-scale helicopters.

General Observations

All on-axis responses attain a coherence close to unity over most of the frequency range where the relevant dynamical effects take place. The generally high coherence obtained for the key helicopter responses speaks for the good instrumentation, the successfully performed flight experiments and the dominantly linear behavior of the rotorcraft for the tested operating conditions. These results show that the derived frequency responses successfully capture the dynamics of our small-scale rotorcraft and thus are adequate for the intended frequency-domain identification.

The frequency responses and coherence measures derived from the flight data will aid us in the development of the parameterized identification model. These results provide clues about what dynamic effects are dominant in the different parts of the frequency range, as well as clues on how the dynamics are coupled.

Chapter 3

DEVELOPMENT OF THE IDENTIFICATION MODEL

The goal of our system identification is to achieve the best possible fit of the estimated frequency responses and, ultimately, the vehicle responses, with a model that is physically meaningful and that has an accuracy well suited for high-bandwidth control design applications.

The foundations for the development of the linear parameterized model are equations of motion derived from first principles. Directly linearizing a detailed rotorcraft first-principles model would yield expressions that are too complex to be used for system identification. Such systems are usually of high order and contain a large number of parameters. The information contained in the estimated frequency responses (number of measured inputs and outputs, as well as spectral content) is typically not sufficient to identify such large systems of equations, i.e. the problem is underdetermined.

Choosing the right model structure and parameterization is usually not accomplished in a single step; rather, it requires multiple iterations. The development of the model is incremental: starting with the basic rigid-body equations of motion and extending them with additional dynamics to satisfy the desired level of detail and accuracy. This process is guided by the clues provided by the estimated frequency responses and intermediate identification results.

In the following, we describe the different steps that were followed for the development of the R-50 model. We start with the description of the equations of motion for a rigid-body system and introduce the stability derivatives model. Then, we describe the techniques used to extend such a linear stability derivatives model to higher orders. We will extend it first, to account for the rotor system, for which we will introduce the necessary knowledge about rotary-wing

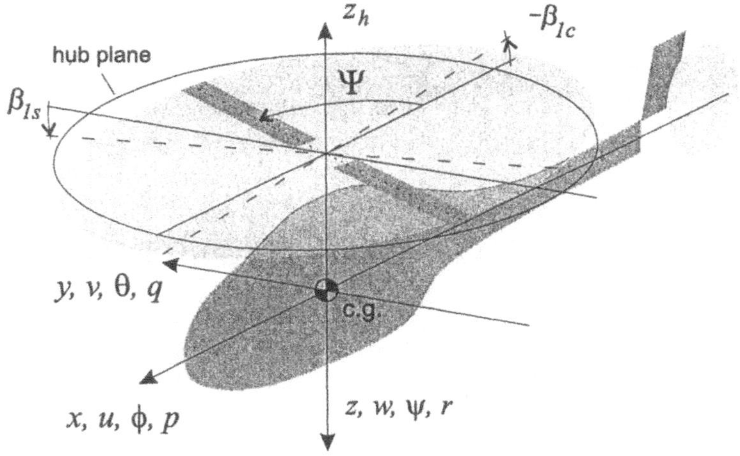

Figure 3.1. **Helicopter with its body-fixed reference frame; origin is at the helicopter center of gravity. The principal variables are shown on the x, y, and z body axes. They include: the body velocities u, v, w; the Euler angles ϕ, θ, ψ; and the body angular rates p, q, r. The main rotor is represented as a disc that can tilt about the rotor hub in the longitudinal and lateral directions. This motion is described through the angles β_{1c} and β_{1s} measured in reference to a plane perpendicular to the rotor shaft (hub plane). The actual rotor blade position is described by the angle Ψ measured from the tail.**

dynamics, second, for the stabilizer bar system. Finally, we explain the yaw and heave dynamics, the identificaton of the actuators, and present the complete paramterized model that will be used in next chapter for the identificaton.

1. Rigid-Body Equations of Motion

Rigid-body equations of motion are used extensively for fixed-wing aircraft. Linear stability derivatives models, which are a linearized form of the rigid-body dynamics, allow for important analysis of the vehicle's dynamics, stability, and control [56]. In rotorcraft, such quasi-steady models are also used for simple flying-qualities analysis [68]. However, as we will see later, for rotorcraft they are not sufficiently accurate to support detailed handling-qualities analysis [30] and high-bandwidth flight-control design [85]. Nevertheless, they represent the natural starting point for our model development. In the following, we will describe the development of the rigid-body equations of motion, and then highlight the main limitation areas.

1.1 Rigid-Body Equations of Motion

The helicopter is a vehicle that is free to simultaneously rotate and translate in all six degrees of freedom (6 DOF). The rigid-body dynamics of such vehicles are described by the Newton-Euler equations of motion [56]. Figure 3.1 shows the helicopter variables in a body-fixed reference frame with origin at the vehicle's center of gravity, or c.g. (axes x, y, z; velocities u, v, w; Euler angles ϕ, θ, ψ; and angular rates p, q, r).

Newton-Euler Equations

The Newton-Euler equations expressed in the inertial reference frame are derived from the principle of conservation of linear and angular momentum. For constant vehicle mass m and moment of inertia (inertial tensor $\boldsymbol{I}$), they are:

$$m\frac{d^I \boldsymbol{v}}{dt} = \boldsymbol{F} \tag{3.1}$$

$$\boldsymbol{I}\frac{d^I \boldsymbol{\omega}}{dt} = \boldsymbol{M} \tag{3.2}$$

where $\boldsymbol{F} = [X \ Y \ Z]^T$ is the vector of external forces acting on the vehicle center of gravity and $\boldsymbol{M} = [L \ M \ N]^T$ is the vector of external moments. In a helicopter, the external forces and moments are produced by: the main rotor and the tail rotors (these forces represent the control forces and moments); the gravitational forces, and the aerodynamic forces produced by the fuselage and the tail surfaces.

Using the kinematic principles of moving reference frames [64], we can express the equations of motion with respect to the body-fixed reference frame:

$$m\dot{\boldsymbol{v}} + m(\boldsymbol{\omega} \times \boldsymbol{v}) = \boldsymbol{F} \tag{3.3}$$

$$\boldsymbol{I}\dot{\boldsymbol{\omega}} + (\boldsymbol{\omega} \times \boldsymbol{I}\boldsymbol{\omega}) = \boldsymbol{M}. \tag{3.4}$$

For the helicopter, $\boldsymbol{v} = [u \ v \ w]^T$ and $\boldsymbol{\omega} = [p \ q \ r]^T$ are the fuselage velocities and angular rates in the body-fixed frame, respectively. For a 6 DOF rigid-body system, Eq. 3.3 produces the three differential equations describing the helicopter's translational motion about its three reference axes

$$\dot{u} = (-wq + vr) + X/m \tag{3.5}$$

$$\dot{v} = (-ur + wp) + Y/m \tag{3.6}$$

$$\dot{w} = (-vp + uq) + Z/m. \tag{3.7}$$

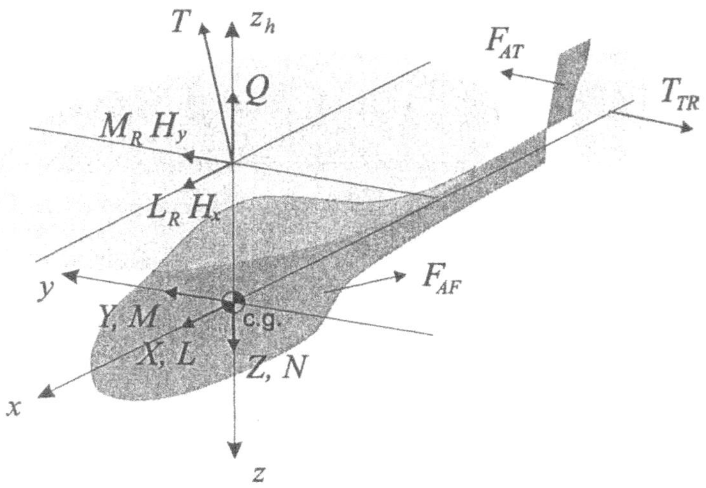

Figure 3.2. Forces and moments acting on the helicopter. The components of the resultant forces and moments acting at the helicopter c.g. are shown as X, Y, Z and L, M, N, respectively. Also shown are the key forces that contribute to the helicopter motion, including: the rotor thrust T; the longitudinal and lateral rotor moments acting on the hub L_R and M_R; the main rotor torque Q; the in-plane rotor forces H_x, H_y; the tail rotor thrust (T_{TR}); the fuselage aerodynamic forces (F_{AF}); and the aerodynamic forces from tail surfaces (F_{AT}).

Similarly, Eq. 3.4 produces the three ordinary differential equations describing the vehicle's rotational motion (we make the assumption that the cross-products of inertia are small, i.e. that the principal axes coincide with the axes of the body reference system):

$$\dot{p} = -qr(I_{yy} - I_{zz})/I_{xx} + L/I_{xx} \tag{3.8}$$

$$\dot{q} = -pr(I_{zz} - I_{xx})/I_{yy} + M/I_{yy} \tag{3.9}$$

$$\dot{r} = -pq(I_{xx} - I_{yy})/I_{zz} + N/I_{zz}. \tag{3.10}$$

These six first-order differential equations describe the rigid-body motion of the vehicle as a function of the external forces and moments. Figure 3.2 shows the components of the forces and moments acting on a helicopter. The next step in the modeling consists of the derivation of mathematical expressions describing these external forces and moments as a function of the control inputs and the vehicle states. In their general form, the vehicle equations of motion, Eqs. 3.5 to 3.10, can be represented as a nonlinear vector differential equation

$$\dot{\boldsymbol{x}} = \boldsymbol{f}(\boldsymbol{x}, \boldsymbol{u}) \tag{3.11}$$

where $\boldsymbol{x}$ is the vehicle state vector, and $\boldsymbol{u}$ is the control input vector. For the rigid-body dynamics, the state vector is

$$\boldsymbol{x} = [u, v, w, \phi, \theta, \psi, p, q, r]^T. \tag{3.12}$$

The basic helicopter control variables are: the lateral and longitudinal cyclic rotor controls $\boldsymbol{u}_{lat}, \boldsymbol{u}_{lon}$; the collective pitch $\boldsymbol{u}_{col}$; and the tail rotor collective control input $\boldsymbol{u}_{ped}$. The control input vector is

$$\boldsymbol{u} = [u_{lat}, u_{lon}, u_{col}, u_{ped}]^T. \tag{3.13}$$

In a detailed first-principle modeling approach, every component and physical effect of the vehicle dynamics are modeled. A detailed nonlinear model usually has several states in addition to the rigid-body ones, resulting in high-order expressions for the equation of motion Eq. 3.11. Additional states are necessary to model, for example, the main rotor, the engine-drivetrain, and effects such as the rotor inflow dynamics. The parameterized identification model is usually derived from linearized rigid-body equations of motion, and the higher-order dynamics are described by highly simplified expressions.

An external force we can readily model is the graviational force. To do this, we need to be able to describe the vehicle orientation with respect to and inertial system.

Euler Orientation Angles

The angular orientation of the aircraft is described by Euler angles. The Euler angles refer to a specific sequence of rotations about the vehicle body axes: the standard sequence for aircraft is yaw (angle ψ about the z axis), pitch (angle θ about the "new" y axis), and roll (angle ϕ about the "new" x axis). The rotation matrix for the compound rotations relates the coordinates of a point given in the inertial reference frame to the point's coordinates in the rotated-body reference frame. The matrix is (using the shorthand notation c for cosine and s for sine):

$$ {}^b_i\boldsymbol{R} = \begin{bmatrix} c\theta c\psi & c\theta s\psi & -s\theta \\ s\phi s\theta c\psi - c\phi s\psi & s\phi s\theta s\psi + c\phi c\psi & s\phi c\theta \\ c\phi s\theta c\psi + s\phi s\psi & c\phi s\theta s\psi - s\phi c\psi & c\phi c\theta \end{bmatrix}. \tag{3.14}$$

Using the rotation matrix, we can determine the components of the gravitational force in the x, y, and z body coordinates. The gravity vector in the inertial reference frame has the components: ${}^i\boldsymbol{g} = [0,\ 0,\ g]^T$, hence the components in the body axes are:

$$ {}^{b}\boldsymbol{g} = {}^{b}_{i}\boldsymbol{R}\,{}^{i}\boldsymbol{g} = \begin{bmatrix} -g\sin\theta \\ g\cos\theta\sin\phi \\ g\cos\theta\cos\phi \end{bmatrix} . \tag{3.15} $$

Kinematics of Angular Motion

The vehicle's angular velocity $\boldsymbol{\omega}$ is related to the time rate of change of the Euler angles through the following relation [64]:

$$ {}^{b}\boldsymbol{\omega} = \begin{bmatrix} p \\ q \\ r \end{bmatrix} = \begin{bmatrix} \dot{\phi} - \dot{\psi}\sin\theta \\ \dot{\theta}\cos\phi + \dot{\psi}\sin\phi\cos\theta \\ \dot{\psi}\cos\phi\cos\theta - \dot{\theta}\sin\phi \end{bmatrix} . \tag{3.16} $$

1.2 Rigid-Body Stability Derivatives Model

The stability derivatives model is a linearized form of the equations of motions where the external forces and moments are represented through the product of derivatives and the rigid-body vehicle states and control inputs. In the following, we first introduce the linearization of the rigid-body equations of motion and then the calculation of the derivatives from a Taylor series expansion.

Linearization of the Equations of Motion

The nonlinear vector differential equation of motion in Eq. 3.11 can be linearized about an equilibrium state $\boldsymbol{x}_0$, i.e.

$$ \dot{\boldsymbol{x}} = \left(\frac{\partial \boldsymbol{f}}{\partial \boldsymbol{x}}\right)_{\boldsymbol{x}_0,\boldsymbol{u}_0} \delta\boldsymbol{x} + \left(\frac{\partial \boldsymbol{f}}{\partial \boldsymbol{u}}\right)_{\boldsymbol{x}_0,\boldsymbol{u}_0} \delta\boldsymbol{u}(t). \tag{3.17} $$

This expression can be presented in the familiar state-space form with system and control matrix $\boldsymbol{A}$ and $\boldsymbol{B}$, respectively

$$ \delta\dot{\boldsymbol{x}} = \boldsymbol{A}\delta\boldsymbol{x} + \boldsymbol{B}\delta\boldsymbol{u} \tag{3.18} $$

which use the linear perturbations of the state and input vectors shown in Eqs. 3.12 and 3.13. The vehicle-state trajectory and control input about the equilibrium point are given by

$$ \boldsymbol{x} = \boldsymbol{x}_0 + \delta\boldsymbol{x} \qquad \boldsymbol{u} = \boldsymbol{u}_0 + \delta\boldsymbol{u}. \tag{3.19} $$

The trim conditions for the chosen operating point are determined by fixing the known states to the values that characterize the equilibrium point and by solving the equilibrium equations $\boldsymbol{f}(\boldsymbol{x},\boldsymbol{u}) = 0$ for the values of the remaining vehicle states and control inputs. A trivial but physically meaningful equilibrium point for a helicopter is the hover flight condition defined by zero velocity and angular velocity

$$\boldsymbol{v}_0 = \boldsymbol{\omega}_0 = [0, 0, 0]^T. \tag{3.20}$$

Other basic equilibrium conditions are the family of level rectilinear trajectories, more complex ones are the family of steady turns and helices.

The linearized translational and rotational rigid-body equations of motion (at some general operating point) are

$$\delta\dot{u} = (-w_0\delta q + \delta w q_0 + v_0\delta r + \delta v r_0) + \Delta X/m \tag{3.21}$$

$$\delta\dot{v} = (-u_0\delta r + \delta u r_0 + w_0\delta p + \delta w p_0) + \Delta Y/m \tag{3.22}$$

$$\delta\dot{w} = (-v_0\delta p + \delta v p_0 + u_0\delta q + \delta u q_0) + \Delta Z/m \tag{3.23}$$

$$\delta\dot{p} = (-q_0\delta r - \delta q r_0)(I_{yy} - I_{zz})/I_{xx} + \Delta L/I_{xx} \tag{3.24}$$

$$\delta\dot{q} = (-p_0\delta r - \delta p r_0)(I_{zz} - I_{xx})/I_{yy} + \Delta M/I_{yy} \tag{3.25}$$

$$\delta\dot{r} = (-p_0\delta q - \delta p q_0)(I_{xx} - I_{yy})/I_{zz} + \Delta N/I_{zz}. \tag{3.26}$$

They describe the vehicle's response about the trim condition as a function of perturbations in the external forces and moments, $\Delta\boldsymbol{F}$ and $\Delta\boldsymbol{M}$, respectively.

Expanding the External Forces and Moments

The external forces and moments can be assumed to be continuous functions of the vehicle's states and control variables. Hence, we can express their perturbation using a Taylor series expansion. Since we want a linear dependancy on the states and controls, we only use the first order terms of the series. For example, for the longitudinal force component we get

$$\Delta X = \frac{\partial X}{\partial u}\delta u + \frac{\partial X}{\partial v}\delta v + \frac{\partial X}{\partial p}\delta p + \cdots + \frac{\partial X}{\partial \delta_{lat}}\delta_{lat} + \frac{\partial X}{\partial \delta_{lon}}\delta_{lon} + \cdots. \tag{3.27}$$

The partial derivatives of the forces (or moments) with respect to the vehicle states are termed stability derivatives. For example, the derivative of the longitudinal force with respect to the longitudinal velocity is abbreviated as:

$$\frac{\partial X}{\partial u} = X_u. \tag{3.28}$$

The derivatives of the forces (or moments) with respect to the vehicle inputs are termed control derivatives. For example, the derivative of the longitudinal force with respect to the lateral cyclic input is abbreviated as:

$$\frac{\partial X}{\partial \delta_{lat}} = X_{\delta_{lat}}. \tag{3.29}$$

Using the stability derivatives notation, the total longitudinal component of the force perturbation is:

$$\Delta X = X_u \delta u + X_v \delta v + X_p \delta p + \cdots + X_{\delta_{lat}} \delta_{lat} + X_{\delta_{lon}} \delta_{lon} + \cdots . \tag{3.30}$$

The same applies to the other components of the force and moment perturbations. For example, for the roll moment perturbation we have:

$$\Delta L = L_u \delta u + L_v \delta v + L_p \delta p + L_q \delta q + \cdots + L_{\delta_{lat}} \delta_{lat} + L_{\delta_{lon}} \delta_{lon} + \cdots . \tag{3.31}$$

Not all vehicle states and input variables will play into each force or moment component. An important task in the development of the model structure consists of deciding which derivatives are relevant in each force or moment. Analytical expressions for the derivatives can be found in rotorcraft literature [74, 42, 68]. These expressions are usually derived by analytical differentiation of detailed nonlinear equations of motion and discarding the term with smaller order of magnitude. Such details can sometimes be helpful in determining which derivatives play a significant role in a particular force or moment.

To simplify the notation in the rest of the book, we will drop the deltas (δ and Δ) from all the variables except the control inputs.

1.3 Limitations of the Rigid-Body Model

In rotorcraft, control forces and moments are produced by the main rotor and the tail rotor. In a rigid-body representation, these forces are described by control derivatives such as $L_{\delta_{lat}}$ for the roll moment produced by the lateral cyclic input. This representation implies that the rotor forces and moment are instantaneous, and proportional to the control inputs.

In reality, the rotor is a dynamic system that responds to both the rotor cyclic and collective controls and the vehicle state (e.g. body angular motion and airspeed). Therefore the rotor forces and moments are not instantaneous, but will depend on the dynamic response of the rotor to the control inputs and vehicle motion. The dependance of the rotor forces and moments on the vehicle motion, and vice-versa, results in a dynamic coupling between these two systems.

The coupled rotor-fuselage dynamics are one of the key characteristics of rotorcraft dynamics. The effects of this coupling are already apparent just beyond the quasi-steady dynamics and omitting to model it in full scale rotorcraft is known to result in poor high-bandwidth fidelity [33]. The estimated roll and pitch frequency responses for the R-50 shown in Chapter 2 did clearly exhibit the characteristics of a second order system. Such characteristics are the trademarks of such coupled dynamics. This issue will be clearer in Chapter 5 where we will illustrate the effect of using rigid-body dynamics on the attitude rate frequency responses.

1.4 Extension of the Rigid-Body Model

To improve the rigid-body model fidelity, higher-order effects that are critical to the vehicle dynamics can be modeled explicitly and then coupled to the rigid-body dynamics. A popular extension that was introduced in full-scale rotorcraft for attitude dynamics analysis [36] and later used with system identification [87], is the "hybrid-model". This model model is essentially a rigid-body model augmented with simplified rotor dynamics. Model extension is not limited to the rotor dynamics; other augmentations found in full-scale rotorcraft include engine-drivetrain and inflow dynamics [23, 89].

Determining what dynamics are critical for small-scale rotorcraft is one of the key tasks in the development of the parameterized identification model. From the frequency responses it is clear that the coupled rotor-fuselage dynamics are critically important for small-scale vehicles. In the following, we will describe the derivation of such coupled rotor-fuselage equations of motion. In Section 4 we will also describe how the stabilizer bar dynamics and the active yaw-damping system, which are often used in small-scale rotorcraft to ease the pilot's workload, are modeled and appended to the fuselage-rotor model.

Extending the rigid-body model by explicitly accounting for specific effects, and formulating those based on first principles, we will also increase the physical meaning of the parameters that will eventually be identified. In the plain rigid-body stability derivatives model, the unmodeled dynamics, such as the rotor or stabilizer bar dynamics, are absorbed by the rigid-body derivatives (see Chapter 5). By explicitly accounting for higher-order dynamics, the identified parameters will gain in physical meaning.

2. Simplified Rotor Dynamics

To account for the rotor dynamics in the stability-derivative model, we need simplified expressions for the rotor equations of motion . With these simplified dynamics, we will then be able to express the rotor forces and moments as a function of the rotor states, and couple the two dynamic systems.

Detailed rotor equations of motion can be extremely complex. The rotor is itself a multi-body system composed of blades that can flex and move independently about hinges or flexible roots. The blades are airfoils with an adjustable pitch angle. Their motions are induced by the changing airloads which follow control inputs, but also the airloads and inertial forces induced by the vehicle motion itself. In addition, a rotor operates in its own wake, resulting in complex aerodynamics conditions. For example, the rotor aerodynamic working state is different for each operating condition (e.g. hover flight or cruise flight), the continuously changing orientation of the blades with respect to the aircraft's direction of motion results in inflow nonuniformities, and the rotor inflow dynamics are responsible for additional dynamic effects.

Not all the above effects are relevant for the type of linear model we need for identification. In the following we will describe the most fundamental aspects of rotor dynamics and summarize the development of a so-called tip-path-plane rotor model used in low-complexity simulation systems and which represents the foundation of the highly simplified model used for the identification.

2.1 Rotor Mechanization and Aerodynamics

Before going into the rotor dynamics, we first describe the blade motion, rotor mechanization and basic rotor aerodynamics.

Blade Motion

Rotor blades are attached to the spinning shaft via a rotor head. Besides its rotation around the hub (speed Ω, position Ψ), the blade can also rotate about three degrees of freedom (DOF). Figure 3.3 shows a schematic view of the three blade DOF. Feathering is the motion of the blade about its length and is described by the blade pitch angle Θ; flapping is the motion of the blade in a direction normal to the rotor disc and is described by the flapping angle β; and lead-lagging is the motion of the blade in the rotor disc plane and is described by the angle ξ.

Significant variations in the design of helicopter rotor heads exist. Blade motion is enabled by mechanical hinges near the blade root (articulated ro-

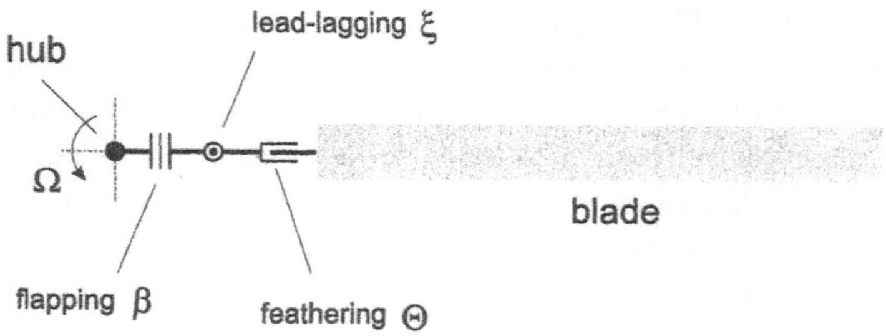

Figure 3.3. Schematic view of the three degrees of freedom of the blade. Each DOF is depicted conceptually by a hinge.

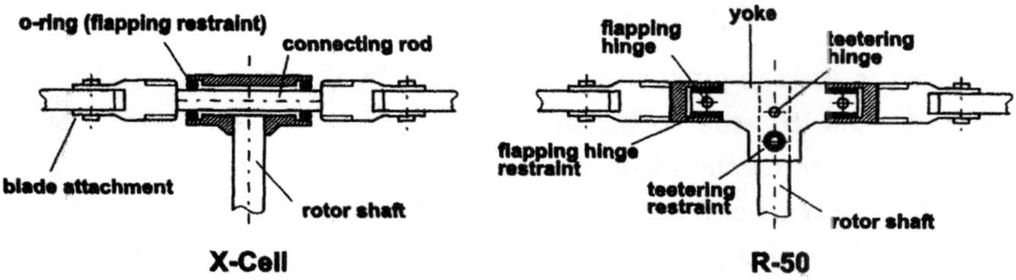

Figure 3.4. Rotor head mechanization for the X-Cell and R-50.

tor), by blade root compliance (hingeless rotors), or by a combination of both. Hybrid head designs are used for the Yamaha R-50's and the X-Cell's rotor (shown in Figure 3.4). In the R-50, the blade flapping takes place about both a flapping hinge and a teetering hinge (tri-hinge system). The motion about the three hinges is restrained by elastomer fittings that act both as springs and dampers. The spring effect is used to transform the blade flapping motion in a hub moment. A stiffer rotor head will be capable of producing a higher moment per unit of blade flapping deflection. The damping influences the phasing of the flapping response (angular deplacement from the maximum aerodynamic force to the maximum flapping deflection). For the X-Cell, there are no actual mechanical flapping hinges. The flapping motion takes place when the rod connecting the two blade roots compresses the elastomer O-rings at each end of the cylindrical cavity. The X-Cell rotorhead is also significantly stiffer than that of the R-50 and produces greater moments.

Swashplate Mechanism

In most rotorcraft, the rotorspeed is kept constant by an electronic governor. The thrust and rotor moments are produced by changing the blade pitch angle. The blade pitch control system is based on a swashplate mechanism. The purpose of this mechanism is to vary the pitch of the blade both in magnitude but also as a function of the blade angular position around the hub. Using the longitudinal and lateral cyclic control inputs, the pilot can control how much the blade pitch changes longitudinally and laterally as the blade rotates. Using the collective control input, the pilot controls the average blade pitch angle. The blade pitch angle as a function of its position Ψ around the hub (the blade angle Ψ is zero when above the tail, and we assume that the blade rotates counter clockwise when seen from above; see Fig. 3.1) is described by

$$\Theta(\Psi) = \Theta_0 - A_1 \cos\Psi - B_1 \sin\Psi \tag{3.32}$$

where Θ_0 is the average pitch angle, which is set by the collective control input δ_{col}; and A_1 and B_1, the coefficients of the cosine and sine terms, are the amount of blade pitch the blade undergoes when it is located above the tail ($-x$ – body direction) and on the right-hand side (y – direction), respectively. They are functions of the longitudinal and lateral cyclic controls, δ_{lon} and δ_{lat}, respectively. We can write them as a function of linear gearing coefficients which transform the pilot stick input into angular blade-root pitch change

$$A_1 = B_{lat}\delta_{lat} \qquad B_1 = A_{lon}\delta_{lon}. \tag{3.33}$$

All four control inputs are given in percentage of the maximum range ($\pm 1 \equiv \pm 100\%$), hence the units for A_{lon} and B_{lat} are $rad/\%$. We chose the sign convention so that the cyclic inputs result in steady angular flapping angles (see Eq. 3.62) that are consistent with the body attitude angles (see Fig. 3.1). As we will see soon, this requires a 90 degrees rotation of the two cyclic inputs, from their corresponding body axes.

Basic Rotor Aerodynamics

To determine the aerodynamic forces acting on the blades, the air velocity at a blade element must first be determined. The velocity at the blade element is represented by a tangential and perpendicular component U_T and U_P, respectively. These two directions are defined with respect to the hub plane (see Fig. 3.1). Figure 3.5 shows an element of the rotor blade airfoil with the velocity components and the angles that will be used to describe the rotor aerodynamics.

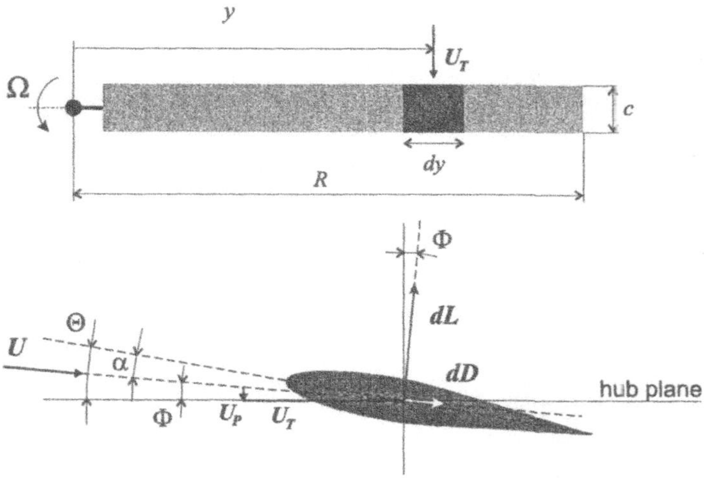

Figure 3.5. **Element of blade airfoil located at a station y along the blade, with blade velocity components, aerodynamic angles, and elemental lift and drag forces.**

The main contribution to the airspeed at the blade is the rotor rotation about the hub. The other contributions are from the vehicle translational and rotational motion, the rotor inflow, and the blade flapping motion. When the rotorcraft moves through the atmosphere, the advancing blade sees a higher airspeed than the retreating blade, resulting in a periodicaly varying airspeed at the blade.

The vehicle speed relative to the undisturbed air is represented by a freestream velocity U_∞ and the angle α_D between the airspeed vector and the hub plane (we assume that the vehicle is oriented into the airstream, i.e. no side slip). Figure 3.6 shows the main components contributing to the total velocity at the blade. The tangential velocity component at a blade station can be written y can be written as

$$U_T = \Omega y + (U_\infty \cos \alpha_D) \sin \Psi. \tag{3.34}$$

The perpendicular velocity component is made of the rotor inflow velocity v_i, which comes as a direct effect of the production of thrust by the rotor (and is given by momentum theory [29]), the perpendicular component of the freestream velocity U_∞, the vehicle rotation (angular rates p and q), and the blade flapping rate:

$$U_P = U_\infty \sin \alpha_D + v_i - y\,(p \sin \psi + q \cos \psi) + y\dot{\beta}. \tag{3.35}$$

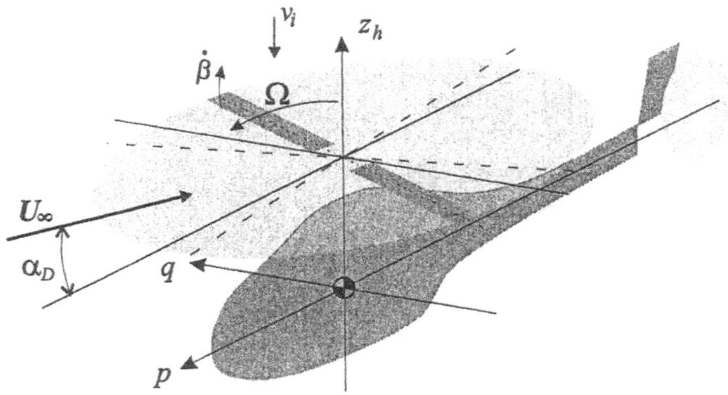

Figure 3.6. Freestream velocity U_∞ making at angle α_D with the hub plane with the main component contributing to the total velocity at the blade; the helicopter rotor speed is Ω, the rotor inflow velocity v_i, the fuselage roll p and pitch q rates, and the blade flapping rate $\dot{\beta}$.

The magnitude of the resultant velocity at the blade element is

$$U = \sqrt{U_P^2 + U_T^2}. \tag{3.36}$$

An important parameter used later is the advance ratio μ, which is defined as the non-dimensional forward velocity $\mu = U/(\Omega R)$.

The blade aerodynamic angle of attack α (angle of the airfoil relative to the resultant velocity) is a function of both the blade pitch angle setting Θ and the aerodynamic inflow angle Φ:

$$\alpha = \Theta - \Phi. \tag{3.37}$$

The inflow angle Φ is given by:

$$\Phi = \tan^{-1}\left(\frac{U_P}{U_T}\right). \tag{3.38}$$

The incremental lift produced by a blade element acts perpendicularily to the resultant velocity. Using linear aerodynamics [52] we have

$$dL = \frac{1}{2}\rho U^2 c C_{l\alpha} \alpha dy \tag{3.39}$$

c is the blade chord length, ρ is the air density, and $C_{l\alpha}$ is the airfoil's lift curve slope. The elemental blade drag force acts in line with the resultant airspeed, similarily we have

$$dD = \frac{1}{2}\rho U^2 c C_{d\alpha}. \tag{3.40}$$

We can now derive the in-plane and out-of-plane force components acting on the blade element,

$$dF_z = dL\cos\Phi + dD\sin\Phi \approx dL \tag{3.41}$$

$$dF_x = dL\sin\Phi + dD\cos\Phi \approx dD. \tag{3.42}$$

The total lift and drag forces acting on one blade is obtained by integrating the elemental forces along the blade length.

We see that as the helicopter is moving through the air, or when cyclic control is applied, the rotating blade is subject to periodic aerodynamic forces. As a result of the periodic aerodynamic forces, the blade undergoes periodic flapping motion which in turn will produce a torque on the helicopter via the rotor hub. In the following we will derive the equations of motion describing the blade flapping dynamics.

Note that in the above we used a steady and uniform inflow velocity v_i. In reality, the rotor inflow will be a function of the blade lift forces (rotor thrust), hence the aerodynamic environment of the blade is itself a dynamic system. The distribution of the inflow will also be nonuniform and it will change as a function of the vehicle motion. We see from this that detailed rotary-wing aerodynamics rapidly become too complex to be studied analyticaly.

2.2 Simplified Rotor Equation of Motion

To illustrate the derivation of the rotor model used to extend the rigid-body stability derivatives model, we follow the development of a so-called tip-path-plane (TPP) model that can be found in literature [42, 68]. We first introduce the simplifying assumptions that are usually made for such TPP models.

Simplifying Assumptions

The tip-path-plane equations of motion are used in medium-fidelity simulation models such as those described in [11] or [37]. The applications of such models are typically limited to an advance ratio of $\mu \leq 0.3$. The typical approximations made in such models (taken mostly from [10]) are

1 Structural simplifications: rigid blade in bending and torsion and linear blade twist; small flapping angle β.

2 Aerodynamic simplifications: small inflow angle $\Phi = U_P/U_T \ll 1$; ignore reversed rotor flow, compressibility, and blade stall; and ignore inflow dynamics and blade tip losses.

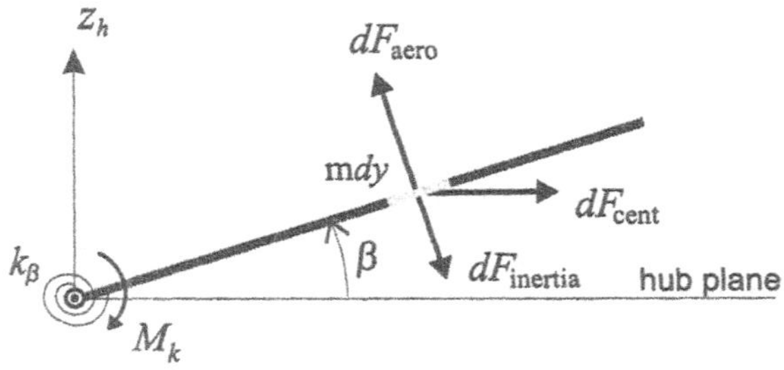

Figure 3.7. **Blade represented as a rigid beam, with flapping angle β. Also shown are the aerodynamic, centrifugal, and intertial forces acting on a blade element of length dy and mass mdy, as well as the moment from the flapping spring acting at root of the blade.**

3 Dynamic simplifications: aircraft motion effects on blade flapping limited to those due to angular accelerations $(\dot{p}, \dot{q})$, angular rates (p, q), and normal accelerations $(\dot{u}, \dot{v}, \dot{w})$.

4 The lead-lag motion, which is mainly the result of the Coriolis forces induced by the flapping motion, produces significantly smaller forces on the hub than the flapping, and will also be ignored.

The blade is represented as a rigid beam and its attachement to the shaft is modeled through a flapping hinge located at the center of the hub (no hinge offset) with a linear torsional spring with spring constant k_β. This center-spring rotor representation can be used to model a variety of hinged and hingeless rotor heads [68]; the effect of a hinge offset and hinge restraint is captured by the spring coefficient.

Derivation of the Rotor Equations of Motion

The equations for the blade flapping motion are derived from the balance of moments about the flapping hinge. Figure 3.7 shows a rotor blade as a rigid beam with the main forces and moments acting on it. They are the moments of the blade aerodynamic force F_{aero}, the centrifugal force F_{cent}, the inertial forces $F_{inertia}$ (resulting from the blade and helicopter motion), and the moment M_k produced by the flapping restraint.

The complete expression for the inertial and aerodynamic forces which will not be included here to keep the expressions concise. They would normally consist of the contributions of the vehicle angular and linear acceleration and the Coriolis acceleration (see [42, 10, 68] for the complete expressions). In the following, we first derive the elemental forces acting on a blade element located at the radial station y. The blade mass m is given per unit length.

For the inertial forces we show the acceleration due to blade flapping $mdy\ddot{\beta}$, and the elemental centrifugal force component normal to the blade element, which is given by $F_{\text{cent}} = mdy\Omega^2 y\beta$.

The aerodynamic forces is dF_z shown in Eq. 3.41. The full expression can be obtained by developing Eq. 3.39. Writing the balance of moments about the flapping hinge, we obtain

$$\int_0^R ydF_z dy - \int_0^R y\left[my\ddot{\beta} - my\Omega^2\beta\right] dy - k_\beta\beta = 0. \tag{3.43}$$

After rearanging the terms in the second integral we get

$$\int_0^R ydF_z dy - \int_0^R my^2 dy\left[\ddot{\beta} + \Omega^2\beta\right] - k_\beta\beta = 0. \tag{3.44}$$

Notice that the second integral is in fact the blade moment of inertia about the flapping hinge, i.e.

$$\int_0^R my^2 dy \stackrel{\text{def}}{=} I_\beta. \tag{3.45}$$

Instead of differentiating with respect to time, we will now differentiate with respect to the angular blade position. With $\Psi = \Omega t$ we get

$$\ddot{\beta} = \Omega^2 \frac{\partial^2\beta}{\partial\Psi^2} = \Omega^2\beta'' \tag{3.46}$$

Using the new notation β'' to describe the differentiation with respect to ψ), and some rearanging, we have

$$I_\beta\Omega^2\left[\beta'' + \beta\right] + k_\beta\beta = \int_0^R ydF_z dy. \tag{3.47}$$

Notice the similarity of this equation to the differential equation of a mass-spring-damper (MSD) system (note that the damping $\dot{\beta}$ term is embedded in the aerodynamic forcing term). If we rearrange the equation in the standard MSD form we obtain

$$\beta'' + \left(1 + \frac{k_\beta}{I_{\beta\Omega^2}}\right)\beta = \frac{1}{I_\beta\Omega^2}\int_0^R ydF_z dy. \tag{3.48}$$

where the coefficient of β is the square of the natural flapping frequency ratio λ_β (relative to the rotor angular speed Ω):

$$\lambda_\beta^2 = 1 + \frac{k_\beta}{\Omega^2 I_\beta} = \frac{\omega_\beta^2}{\Omega^2}. \tag{3.49}$$

For the case where there is no flapping restraint ($k_\beta = 0$, e.g. a teetering rotor), the natural flapping frequency is equal to the rotor angular speed Ω ($\lambda_\beta = 1$). Since the rotor system is a second-order system, and the aerodynamic forces produced by a cyclic blade pitch change (or gust disturbance) will excite the rotor at or near its natural frequency, the maximum deflection of the flapping response will exhibit an approximately 90 degree phase shift from the location where it experienced the maximum aerodynamic force. For a free teetering rotor ($k_\beta = 0$) the aerodynamic forces occur at exactly the natural frequency and the phase shift is exactly 90 degrees. In the presence of a flapping restraint (or hinge offset), we see from Eq. 3.49 that the natural flapping frequency will be higher than the rotor angular speed ($\lambda_\beta > 1$). Thus, in this case, the response phase shift will be less than 90 degrees. An adequate phasing of the swashplate gearing (input pre-phasing) usually takes care of this effect.

Rotor Tip-Path-Plane Equation

The flapping motion is a 2π–periodic function. Thus the general solution to the flapping equation can be expressed as a the Fourier series [29]

$$\beta(\Psi) = \beta_0 - \beta_{1c}\cos\Psi - \beta_{1s}\sin\Psi - \beta_{2c}\cos 2\Psi - \beta_{2s}\sin 2\Psi - \cdots\ . \qquad (3.50)$$

In full-scale helicopters, the magnitude of the second harmonic are less than 10% of the magnitude of the first harmonic [29]. Hence, the second and higher harmonics in the Fourier series can be ignored, i.e.

$$\beta(\Psi) \approx \beta_0(t) - \beta_{1c}(t)\cos\Psi - \beta_{1s}(t)\sin\Psi. \qquad (3.51)$$

This first-harmonic representation of the blade flapping motion defines the rotor tip-path-plane (TPP) equation,. This type of motion results in a cone-shaped rotor like the one illustrated Figure 3.8. The top of the cone is the TPP. The non-periodic term β_0 describes the coning angle and the coefficients of the first harmonic β_{1c} and β_{1s} describe the tilting of the rotor tip-path-plane in the longitudinal and lateral directions, respectively. In the following, we will use the notation: a instead of β_{1c}, b instead of β_{1s}, and a_0 instead of β_0.

The flapping equation of motion (Eq. 3.48) with the tip-path-plane solution reduces to the tip-path-plane equation of motion. To complete the transformation, we substitute the tip-path-plane expression Eq. 3.51 into the flapping differential equation Eq. 3.48; match all the non-periodic terms with a_0, the sine terms with b, and the cosine terms with a; we define the tip-path-plane state vector $\mathbf{a} = [a_0, a, b]^T$; and use the expression from Eq. 3.32 for the blade pitch

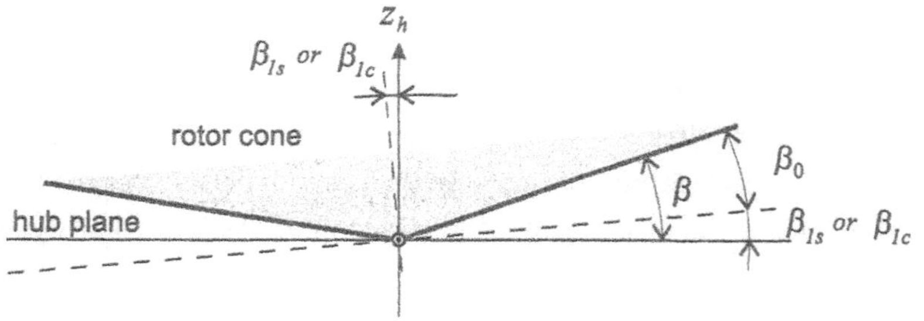

Figure 3.8. Tip-path-plane rotor representation. Flapping angle β is represented by a Fourier series, with non-periodic term β_0 describing the coning angle and the first harmonics β_{1s} (or β_{1c}) describing the lateral (longitudinal) tilting of the tip-path-plane angle.

angle. Through these operations, we obtain a second-order matrix differential equation:

$$\ddot{\mathbf{a}} + \boldsymbol{D}\dot{\mathbf{a}} + \boldsymbol{K}\mathbf{a} = \boldsymbol{F} \tag{3.52}$$

where $\boldsymbol{D}$ is the damping matrix, $\boldsymbol{K}$ is the stiffness matrix, and $\boldsymbol{F}$ is the forcing term.

The complete equations can be found in [10]. We will present the simplified rotor model used to extend the rigid-body stability derivatives model next, starting with a brief description of the tip-path-plane dynamic modes. Looking at the modes will help us make further simplifications to the flapping equation and finally get the highly simplified rotor equations that we will use in our identification model.

Tip-Path-Plane Dynamic Modes

If we were to account for coning, the tip-path-plane rotor dynamics have three oscillatory modes of motion: the coning mode, and the advancing and regressing flapping modes [42, 68]. The coning mode is characterized by blade flapping motions that result in a change in the coning angle β_0 (see Fig. 3.8). Its natural frequency is equal to the natural blade flapping frequency λ_β. This mode is typically excited by rotor load variations, such as collective pitch changes. The advancing mode corresponds to a nutation-type motion of the rotor cone. It has a natural frequency of the order of $\lambda_\beta + 1$, i.e. about twice the rotor angular speed. Last, the regressing mode is a precession-type motion of the rotor cone, with a natural frequency of the order of $\lambda_\beta - 1$.

The significance of these modes for the vehicle motion is dermined by how likely they are to couple with the fuselage rigid-body motions. The coning and advancing modes are too fast to couple into the fuselage modes, while the regressive flapping mode has a natural frequency that is close to the natural frequency of the fuselage motion, and hence has a tendency to couple with the fuselage. This is the mode involved in the coupled rotor-fuselage dynamics discussed earlier.

2.3 First Order Tip-Path-Plane Equations

For system identification, only the dynamic effects that pertain to the vehicle dynamics should be included in the identification model. Hence the rotor equations can be further simplified to only account for the low frequency flapping dynamics.

As a starting point we use the $\boldsymbol{D}$, $\boldsymbol{K}$ and $\boldsymbol{F}$ matrices from [10] to which we introduce the following additional simplifications: we discard the effects of translational speed (by setting $\mu = 0$), we set zero pitch-flap coupling coefficient ($K_1 = 0$), and discard the coning dynamics. The damping and stiffness matrices used in [10] reduce to

$$\boldsymbol{D} = \Omega \begin{bmatrix} \frac{\gamma}{8} & 2 \\ -2 & \frac{\gamma}{8} \end{bmatrix}; \quad \boldsymbol{K} = \Omega^2 \begin{bmatrix} \frac{k_\beta}{I_\beta \Omega^2} & \frac{\gamma}{8} \\ -\frac{\gamma}{8} & \frac{k_\beta}{I_\beta \Omega^2} \end{bmatrix} \tag{3.53}$$

and the forcing term becomes

$$\boldsymbol{F} = \Omega^2 \begin{bmatrix} \frac{\gamma}{8} & 0 \\ 0 & \frac{\gamma}{8} \end{bmatrix} \begin{bmatrix} A_1 \\ B_1 \end{bmatrix} + \Omega \begin{bmatrix} -2 & -\frac{\gamma}{8} \\ -\frac{\gamma}{8} & 2 \end{bmatrix} \begin{bmatrix} p \\ q \end{bmatrix} + \Omega \begin{bmatrix} 0 & -\frac{1}{\Omega} \\ -\frac{1}{\Omega} & 0 \end{bmatrix} \begin{bmatrix} \dot{p} \\ \dot{q} \end{bmatrix} \tag{3.54}$$

where A_1 and B_1 are, respectively, the lateral and longitudinal cyclic blade pitch angle (see Eq. 3.32), k_β is the spring constant of the flapping restraint, I_β is the blade moment of inertia about the flapping hinge, and γ is the blade Lock number which represents the ratio between the aerodynamic and inertial forces acting on the blade

$$\gamma = \frac{\rho c C_{l\alpha} R^4}{I_\beta} \tag{3.55}$$

where R is the rotor radius, ρ is the air density, c is the blade chord length, and $C_{l\alpha}$ is the lift curve slope.

Next, we will reduce the order of the rotor equations to discard the high-frequency dynamics. This is achieved by dropping the higher-order terms in the TPP equations of motion Eq. 3.52, i.e. the blade flapping accelerations $\ddot{a}$ and $\ddot{b}$ and the vehicle angular accelerations $\dot{p}$ and $\dot{q}$.

Using equation Eq. 3.52 with the above matrices and simplifications produces the following coupled first-order tip-path-plane equations of motion:

$$\frac{16}{\gamma\Omega}\dot{a} = -a - \frac{16}{\gamma\Omega}q + \frac{p}{\Omega} + \frac{8}{\gamma\Omega^2}\frac{k_\beta}{I_\beta}b - B_1 \qquad (3.56)$$

$$\frac{16}{\gamma\Omega}\dot{b} = -b - \frac{16}{\gamma\Omega}p - \frac{q}{\Omega} - \frac{8}{\gamma\Omega^2}\frac{k_\beta}{I_\beta}a + A_1. \qquad (3.57)$$

These equations are the low-frequency rotor dynamic approximations which capture the key tip-path-plane responses due to control inputs and vehicle motion.

The key terms are:

$$\frac{16}{\gamma\Omega} \qquad (3.58)$$

which corresponds to the rotor time constant τ_f, which is a key parameter of the tip-path-plane dynamics. Notice its dependance on the rotor Lock number and angular speed.

$$-\tau_f q \quad \text{or} \quad -\tau_f p \qquad (3.59)$$

is the longitudinal (lateral) flapping produced by a body pitching rate q (rolling rate p) which corresponds to the important pitch (roll) rotor damping.

$$-\frac{q}{\Omega} \quad \text{or} \quad +\frac{p}{\Omega} \qquad (3.60)$$

is the lateral (longitudinal) flapping resulting from the change of blade angle of attack produced by a body rolling rate p (pitching rate q). This is one of the main rotor cross-coupling effects.

$$\frac{8}{\gamma\Omega^2}\frac{k_\beta}{I_\beta} = \frac{8}{\gamma}(\lambda_\beta^2 - 1) \qquad (3.61)$$

is another cross-coupling effect arising in the presence of a flapping restraint (and hinge offset). Note above that it can be rewritten in term of the natural frequency ratio λ_β, which shows how it relates directly to the phasing of the flapping response. For a teetering rotor $(\lambda_\beta^2 = 1)$ this term would vanish. We represent this term by the derivatives A_b and B_a for the longitudinal and lateral flapping, respectively.

From the first order TPP equations we can determine the steady longitudinal and lateral flapping responses for a step in the cyclic inputs simply by setting $\dot{a} = \dot{b} = 0$ (we assume no fuselage angular motion)

$$a = A_b b - A_{lon}\delta_{lon} \qquad (3.62)$$

$$b = -B_a a + B_{lat}\delta_{lat}. \qquad (3.63)$$

We see that the rotor flaps proportionally to the blade pitch angle. This means that the rotor TPP takes an angular position for which the cyclic pitch produces zero cyclic blade pitch variation (called the plane of zero feathering, see flapping-feathering equivalence [29]).

3. Coupling Rotor and Fuselage Dynamics

To complete the coupling of the rotor tip-path-plane equations of motion with the rigid-body equations of motion, we need to express the forces and moments produced by the rotor in terms of the rotor flapping.

3.1 Rotor Forces and Moments

The forces and moments the blades exert on the rotor head, and ultimately on the fuselage, are usually determined by calculating the resulting effects at the blade attachment. Complete expressions for the rotor forces and moments take into account the detailed aerodynamic loading of the blades and the inertial forces resulting from the blade motion (see [6, 68]). As with the rotor dynamics, details about the bending of the blades (flexible mode shapes) and the blade in-plane motion and torsion are not relevant to our model development for linear system identification.

In the following, we will describe the simplified expressions for the rotor forces and moments used with the tip-path-plane rotor model, which will allow us to couple the rotor and the fuselage equations of motion.

Simplified Rotor Forces and Moments

Figure 3.9 shows the relation between the rotor tip-path-plane angle, the thrust vector, and the forces and moments produced by the rotor.

The rotor hub forces are considered to be the result of the thrust vector tilting. We make the assumption that the direction of the thrust vector stays perpendicular to the rotor tip-path-plane (usually correct for hover and low-speed flight). Following this assumption, the hub forces are the projections of the thrust vector on the x and y hub axes. The thrust components along the vehicle body axes are:

$$T_x = -T \sin a \cos b \approx -Ta \tag{3.64}$$

$$T_y = T \sin b \cos a \approx Tb \tag{3.65}$$

$$T_z = -T \cos a \cos b \approx -T. \tag{3.66}$$

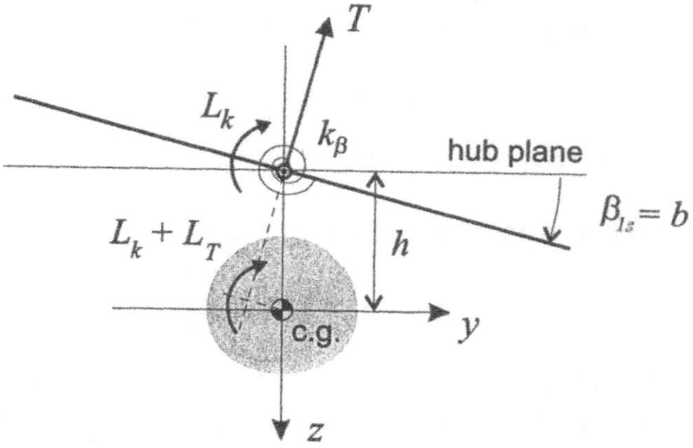

Figure 3.9. **Roll moments resulting from the rotor blade flapping deflection. Two contributions are illustrated: the hub moment resulting from the flapping spring k_β (L_k) (based on a center-spring equivalent hub model) and the moment resulting from the tilting of the thrust vector T (L_T). The orientation of the thrust is simply based on the TPP states.**

The moment produced by rotor flapping acting on the fuselage consists of the moment produced by the tilting of the thrust vector and the rotor hub moments. The moment due to the tilting of the thrust vector is defined from the longitudinal and lateral in-plane thrust components. These forces act on the fuselage with a moment arm of length equal to the hub to fuselage center-of-gravity distance h, resulting in the following roll and pitch moments:

$$L_T = hT_y \approx hTb \tag{3.67}$$

$$M_T = h(-T_x) \approx hTa. \tag{3.68}$$

The rotor hub moments for the center-spring rotor model is simply

$$M_\beta = k_\beta \beta. \tag{3.69}$$

Using the TPP lateral and longitudinal flapping b and a, respectively, the lateral and longitudinal hub torsional moments are

$$L_k = k_\beta b \tag{3.70}$$

$$M_k = k_\beta a. \tag{3.71}$$

The total moment due to rotor flapping acting on the fuselage is obtained by summing the thrust orientation and hub stifness contributions, i.e.

$$L_R = (hT + k_\beta)\, b \tag{3.72}$$
$$M_R = (hT + k_\beta)\, a. \tag{3.73}$$

Hence, we can write a compound spring constant

$$K_\beta = k_\beta + hT. \tag{3.74}$$

Notice how it depends on the rotor thrust. The ratio between moments resulting from blade restraint stiffness and thrust orientation is a distinguishing characteristic between different rotorcraft designs.

Further considerations about this topic will be given at the end of Chapter 4. There we will use these equations to compare the identified values of the rotor moment derivatives with the values predicted by the above expressions.

Rotor Force and Moment Derivatives

We can express the forces and moments using stability derivatives. Note that in the translational and rotational equations, the forces are normalized by the mass of the aircraft, and the moments are normalized by the respective moments of inertia. In hover and slow flight conditions, the thrust is approximately equal to the weight of the vehicle $T = mg$. Hence, we get the following longitudinal and lateral forces derivatives

$$X_a = -T/m = -g \tag{3.75}$$
$$Y_b = T/m = g. \tag{3.76}$$

From Eq. 3.74 we obtain the following pitch and roll moments derivatives

$$M_a = K_\beta/I_{yy} = (k_\beta + hT)/I_{yy} \tag{3.77}$$
$$L_b = K_\beta/I_{xx} = (k_\beta + hT)/I_{xx}. \tag{3.78}$$

3.2 Coupled Rotor-Fuselage Equations of Motion

The rigid-body fuselage equations (Eqs. 3.5 through 3.9) and the rotor tip-path-plane equations (Eqs. 3.56 and 3.57) are coupled through flapping derivatives. In the translational longitudinal and lateral equations of motion (Eqs. 3.79 and 3.80), the input derivatives X_{lon} and Y_{lat} are replaced by the flapping

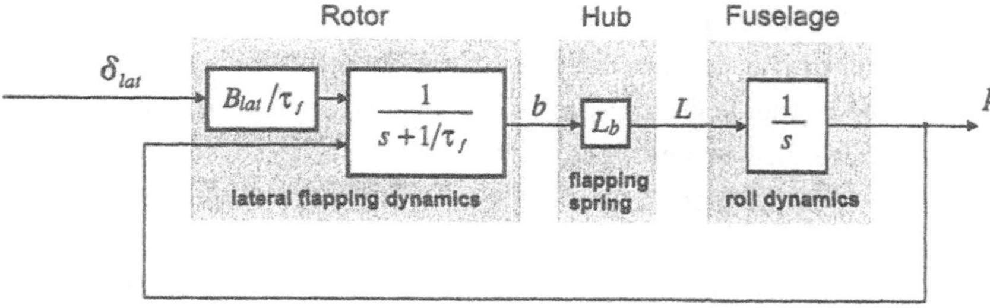

Figure 3.10. **Blockdiagram of the roll dynamics illustrating the coupling between the rotor flapping motion and the fuselage rigid-body dynamics.**

derivatives X_a and Y_b, respectively, i.e.

$$\dot{u} = (-w_0 q + v_0 r) - g\theta + X_u u + \cdots + X_a a \tag{3.79}$$

$$\dot{v} = (-u_0 r + w_0 p) + g\phi + Y_v v + \cdots + Y_b b \tag{3.80}$$

$$\dot{w} = (-v_0 p + u_0 q) + Z_u u + Z_w w + \cdots + Z_{col}\delta_{col}. \tag{3.81}$$

In the roll and pitch equations of motion (Eqs. 3.82 and 3.83) the input derivatives M_{lon} and L_{lat} are replaced by the flapping derivatives M_a and L_b , i.e.

$$\dot{p} = L_u u + L_v v + \cdots + L_b b \tag{3.82}$$

$$\dot{q} = M_u u + M_v v + \cdots + M_a a \tag{3.83}$$

$$\dot{r} = N_r r + \cdots + N_{ped}\delta_{ped}. \tag{3.84}$$

Figure 3.10 shows a block-diagram illustrating the coupling between the rotor and the fuselage for the roll dynamics.

It is important to point out some key differences with the rigid-body equations:

- the rotor forces and moments are represented by the flapping derivatives X_a, Y_b and M_a, L_b, respectively; the cyclic commands δ_{lon} and δ_{lat} now go directly into the rotor dynamics

- The roll and pitch damping derivatives L_p and M_q are dropped, since the roll and pitch damping is now captured by the rotor damping (see Eq. 3.59). For example, for the roll damping, the flapping response to a roll rate p is $-\tau_f p$. Hence, the steady damping moment produced by the roll motion is $-L_b\tau_f$.

- The speed derivatives M_u and L_v are generally retained in the roll and pitch equations of motion. Physically, the speed effect on the angular dynamics occurs mainly at the rotor level (i.e. rotor speed derivatives A_u and B_v), which represent the steady longitudinal and lateral flapping due to a speed perturbation in the longitudinal or lateral direction, respectively. Separating the fuselage and rotor aerodynamic speed effects has been problematic in our system identification.

All the other quasi-steady derivatives are retained to account for the low-frequency dynamics. As a result, the hybrid model has the same characteristics in the low-frequency range as the quasi-steady model, while providing enhanced high-frequency fidelity due to the added rotor dynamics [87]. The key enhancement compared with the quasi-steady model is the ability to capture the coupled rotor-fuselage dynamics.

Note also the centrifugal terms in the translational equations (Eqs. 3.79 to 3.80), which need to be taken into account for flight conditions beyond hover and slow-speed (trim condition $[u_0, v_0, w_0]$).

4. Small-Scale Rotorcraft Model Extensions

Small-scale helicopters are usually fitted with a stabilizer bar and an active yaw damping system. These two systems act as dynamic augmentations and have major effects on the vehicle response. Therefore, in order to allow a physically meaningful identification, it is important to try to model these systems explicitly. Doing so will also make it possible to study the systems' effects on the dynamics. At the same time, we must determine if other model augmentations, such as rotor inflow or engine-drivetrain dynamics, are necessary.

4.1 Typical Features of Small-Scale Rotorcraft

Small-scale helicopters have a higher sensitivity to control inputs (in terms of angular rate attained per unit of control input) than their full-scale counterparts. Because the vehicle dynamics are unstable, the high bandwidth and control sensitivity make it challenging or even impossible for a human to control such a vehicle without some form of stability augmentation. The stabilizer bar acts as a lagged rate feedback in the pitch and roll axes, reducing the bandwidth and control sensitivity to cyclic lateral and longitudinal inputs. The active yaw damping system produces a similar effect on the vehicle yaw dynamics.

Figure 3.11 shows the interconnection of the bare vehicle airframe (fuselage with plain rotor system) with the main subsystems. The stabilizer bar and the active yaw damping system are represented as feedback systems with feedback

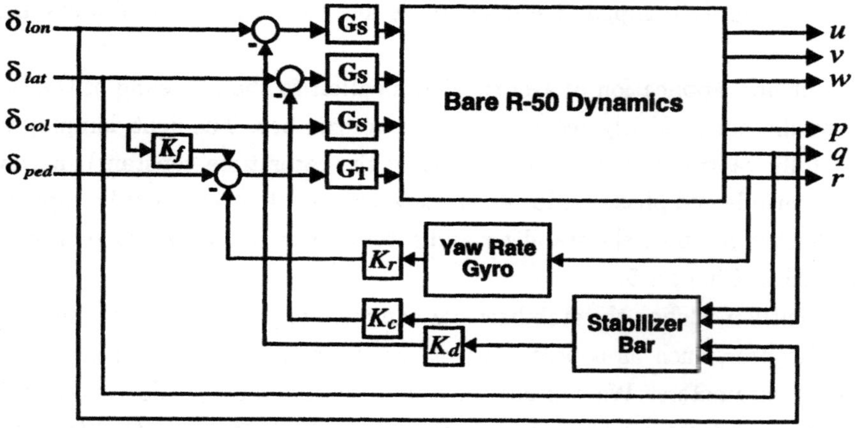

Figure 3.11. **Blockdiagram showing the interconnection between the bare vehicle airframe and the components of the augmented airframe: the actuators, the yaw rate feedback, and the stabilizer bar.**

gains K_c and K_d for the stabilizer bar and K_r for the active yaw damping system. Also shown are the servo actuators: G_S for the three swashplate actuators used for the cyclic and collective control and G_T for the actuator controlling the tail rotor pitch. To reduce the effect of collective input changes on the yaw dynamics, the Yamaha R-50 uses a static crossfeed between the collective input and the pedal input K_f. We refer to the entire system as the "augmented vehicle dynamics".

In the following we will extend the fuselage-rotor model to include all the above features. Explicitly modeling all these subsystems will allow us to analyze their effects. For example, we can study the effect of the stabilizer bar by stripping it off the airframe or varying its key parameters. This also provides for the possibility of treating the parameters of these subsystems as design parameters when developing a control system.

4.2 Coupled Rotor-Stabilizer Equations

The frequency responses of the roll p and pitch q rates to the cyclic lateral δ_{lat} and cyclic longitudinal δ_{lon} inputs (see Figure 2.3) show a pronounced lightly-damped second-order behavior. This characteristic is present for both the on-axis and the off-axis responses. As we just saw, the second-order nature of the response results from a dynamic coupling, namely the coupling between the airframe angular motion and the regressive rotor flap dynamics (blade flapping

a, b). The lightly damped characteristic is due to the presence of a stabilizer bar [35].

In our first application of system identification to the modeling of the R-50 [60], we did not explicitly account for the stabilizer bar. The light damping was succesfully captured (moslty by increasing the rotor time constant), however the magnitude and phase exhibits some key features that could not be captured. The effect of lumping the stabilizer and rotor system in a single system will be discussed in Chapter 5.

Explicitly modeling the stabilizer bar can yield important information about the helicopter, such as a precise identification of the stabilizer's and rotor's physical parameters. We will also be able to determine the effect of changing stabilizer parameters, such as the gearing of the Bell mixer, on the stability of the vehicle. This is particularly important for the stabilizer bar, since it acts as a stability augmentation.

It is important to point out that the stabilizer bar acts as a feedback system, and hence, when identifying a helicopter with a stabilizer bar, the typical limitations of closed-loop identification apply (see Eq. 2.28).

Stabilizer Equations of Motion

The stabilizer bar can be regarded as a secondary rotor. The blades, which consist of two simple paddles mounted at the extremities of a steel rod, are attached to the rotor shaft above the main rotor through an unrestrained teetering hinge. These paddles receive cyclic inputs from the swashplate in a manner similar to that of the main blades, only here no collective pitch is used (the stabilizer bar is not designed to produce thrust, hance it also does not exhibit coning).

The paddle flapping motion can be modeled using the same first-order tip-path-plane equation 3.51 used for the main rotor

$$\beta_{\mathrm{stab}}(\Psi) = -\beta_{\mathrm{stab},1c} \cos\Psi - \beta_{\mathrm{stab},1s} \sin\Psi \tag{3.85}$$

where the stabilizer bar flapping states are defined similarly to the main rotor flapping states. We use the notation d for $= \beta_{1c}$ and c for $= \beta_{1s}$.

The dynamics are also based on the same first-order tip-path-plane equations used for the main rotor (3.56 and 3.57). Note that the term involving k_β vanishes. Even further simplifications can be made.

The stabilizer's Lock number γ_s, is usually significantly smaller than the main blade Lock number. For example, for the R-50, the stabilizer bar Lock number has a theoretical value of $\gamma_s \approx 1.1$ compared to $\gamma \approx 3.95$ for the main blades (X-Cell: $\gamma_s \approx 0.8$ and $\gamma \approx 3.92$). The smaller Lock number results in a larger flapping time constant (Eq. 3.58). It also reduces the coupling between lateral and longitudinal flapping. Namely inspecting the steady pitch flapping response

$$a = -\frac{16}{\gamma\Omega}q + \frac{1}{\Omega}p + A_{lon}\delta_{lon} = \frac{1}{\Omega}\left(p - \frac{16}{\gamma}q\right) + A_{lon}\delta_{lon} \qquad (3.86)$$

with $\gamma \approx 1$ the off-axis p term is more than one order of magnitude smaller than the on-axis q term. Hence we can use the following simplified equations

$$\tau_s\dot{d} = -d - \tau_s p + D_{lat}\delta_{lat} \qquad (3.87)$$

$$\tau_s\dot{c} = -c - \tau_s q + C_{lon}\delta_{lon} \qquad (3.88)$$

where D_{lat} and C_{lon} are, respectively, the lateral and longitudinal stick inputs to the stabilizer bar's cyclic pitch.

Coupling the Stabilizer Bar to the Main Rotor

The stabilizer bar does not produce thrust (it has no collective blade pitch setting), and the paddles are free to teeter (no restraint) about the rotor shaft. Hence, the stabilizer bar does not produce any significant force or moment on the hub. The stabilizer bar affects the vehicle dynamics solely by augmenting the cyclic pitch command to the main rotor blade via the Bell mixer (see Fig. 3.12). The Bell mixer is a mechanical linkage arrangement that augments the main blades' cyclic control inputs with a component that is proportional to the stabilizer bar flapping angle. The resulting augmented lateral and longitudinal main rotor cyclic commands can be written as:

$$\bar{\delta}_{lat} = \delta_{lat} + K_d d \quad \text{and} \quad \bar{\delta}_{lon} = \delta_{lon} + K_c c. \qquad (3.89)$$

The derivatives K_d and K_c represent the gearing of the stabilizer bar mixer and are fixed by the mixer's linkage system.

The main rotor tip-path-plane equations of motion are taken from Eqs. 3.56 and 3.57 but use the augmented cyclic input $\bar{\delta}_{lon}$ and $\bar{\delta}_{lat}$ in the on-axis input channels, i.e.

$$\tau_f\dot{a} = -a - \tau_f q + A_b b + A_{lon}(\delta_{lon} + K_c c) + A_{lat}\delta_{lat} \qquad (3.90)$$

$$\tau_f\dot{b} = -b - \tau_f p + B_a a + B_{lat}(\delta_{lat} + K_d d) + B_{lon}\delta_{lon}. \qquad (3.91)$$

Similar to the stabilizer bar, we dropped the off-axis flapping response terms due to vehicle angular motion. However, we added input cross-coupling derivatives $A_{lat}\delta_{lat}$ and $B_{lon}\delta_{lon}$ to capture possible unmodelled off-axes effects.

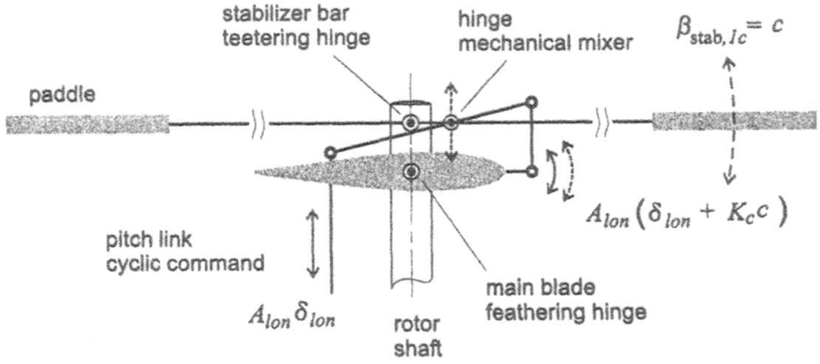

Figure 3.12. **Rotor-stabilizer mechanization. The main blade pitch control receives both the cyclic pitch command and a component imposed by the stabilizer bar that is proportional to the stabilizer bar flapping angle. The cyclic pitch augmentation is implemented through a mechanical mixer.**

By expanding the terms in brackets in Eqs. 3.91 and 3.90 we define the stabilizer bar coupling derivatives $B_d = B_{lat}K_d$ and $A_c = A_{lon}K_c$. The relations between the derivatives and the gearing of the Bell mixer are:

$$K_d = \frac{B_d}{B_{lat}} \quad \text{and} \quad K_c = \frac{A_c}{A_{lon}}. \tag{3.92}$$

In reality, since the Bell mixer operates the same way independently of the blade position around the hub, the gearing is the same for both axes. The relations of Eq. 3.92 could be used as a constraint between the derivatives B_{lat} and B_d (A_{lon} and A_c) to reduce the number of unknown parameters during identification. For the identification of the R-50, we were able to leave both derivatives free and obtained similar values.

4.3 Yaw Dynamics

The equations of motion for the yaw dynamics are derived from the Euler equation (Eq. 3.10). The principal contribution to the external yaw moment results from the moment of the tail rotor thrust. The tail rotor thrust is controlled by the blade pitch control δ_{ped}. Due to its small size and fast rotor speed, the thrust response to a blade pitch change can be considered quasi-instantaneous. It is much faster than the fuselage yaw dynamics and hence no tail rotor dynamics need to be modeled. A time delay is usually sufficient to account for the unmodeled dynamics. Airspeed also produces moments, both through the vertical tail aerodynamic forces and through changes in the tail rotor thrust (translational lift effects). Such effects are usually modeled through

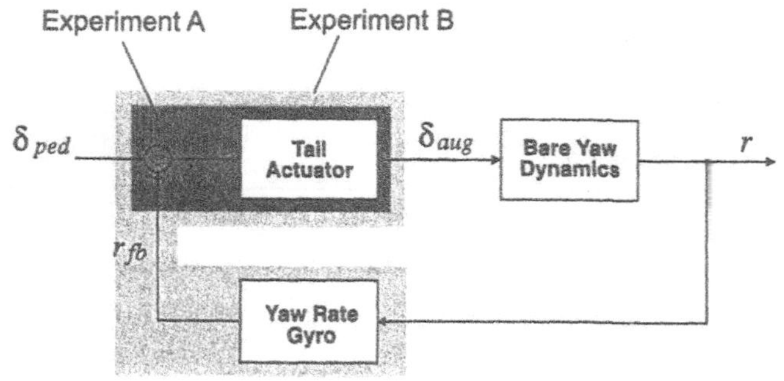

Figure 3.13. **Blockdiagram of the augmented yaw dynamics (active yaw damping system). The closed loop comprises: the helicopter yaw dynamics, the yaw rate gyro, and the tail rotor blade pitch actuator. The shaded boxes indicate the parts of the system that where identified during ground experiments.**

speed derivatives N_u and N_v. Finally, the tail boom and the tail rotor produce damping moments that can be modeled through the stability derivative N_r. The baseline yaw stability derivatives model is

$$\dot{r} = N_v v + N_r r + N_{ped}\delta_{ped}. \tag{3.93}$$

The corresponding transfer function for the yaw response to the tail rotor control is first-order:

$$\frac{r}{\delta_{ped}} = \frac{N_{ped}}{s - N_r}. \tag{3.94}$$

The estimated frequency response for the yaw rate response (see Figure 3.14), however, showed second-order characteristics. The higher order dynamics were attributed to the presence of an active yaw damping system, which is enabled for all piloted flights. To capture the actual dynamics of this system, we need to explicitly account for this dynamic augmentation.

Modeling the Active Yaw Damping System

The yaw damping system is realized through negative feedback of the helicopter heading rate, which is provided by an angular rate gyro. Figure 3.13 shows the block-diagram of the active yaw damping system. It includes the helicopter yaw dynamics, the yaw rate gyro, and the actuator that is used for the implementation of the collective pitch command, which controls the tail rotor thrust.

From the flight experiment, only the measurements of the pilot "pedal" input δ_{ped} and the helicopter yaw rate r are available. Identifying all the sub-elements

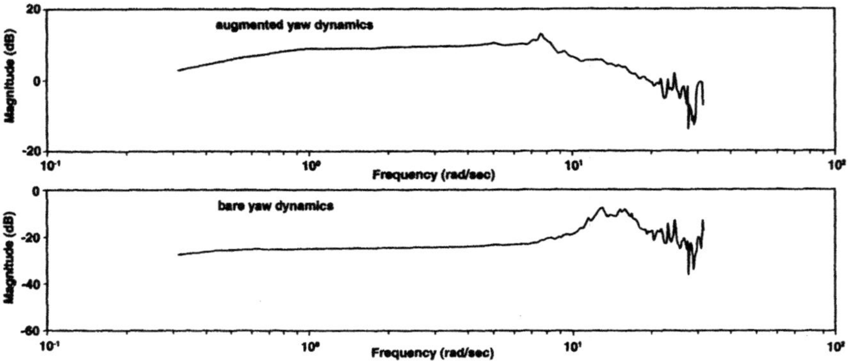

Figure 3.14. Frequency responses of the augmented yaw dynamics (closed-loop dynamics including yaw gyro and actuator) obtained from the flight data; and frequency responses of the bare yaw dynamics, extracted from the bare dynamics using frequency-response arithmetic.

of the augmented yaw dynamics from only these measurements is not feasible. However, the tail rotor actuator and the yaw rate gyro can be identified individually from ground experiments.

Component-Wise Identification

For the separate identification of the tail rotor actuator, the actuator deflection was measured using an encoder, while the tail rotor control input (δ_{ped}) was moved using the frequency-sweep technique (Figure 3.13, Experiment A). From the recorded input-output data, the actuator frequency response was calculated to be (H_{act}). For the yaw rate gyro, the helicopter was swung about the yaw axis following frequency-sweep type motion. Here again, the actuator deflection was measured using the encoder, and the yaw rate was measured using the on-board instrumentation (Figure 3.13, Experiment B). From the recorded input-output data, we calculated the frequency response of the combined rate gyro and tail actuator system ($H_{\text{gyro+act}}$). The frequency response for the rate gyro H_{gyro} can be extracted from $H_{\text{gyro+act}}$ using frequency-response arithmetic

$$H_{\text{gyro+act}} = H_{\text{act}} H_{\text{gyro}} \;\Rightarrow\; H_{\text{gyro}} = H_{\text{act}}^{-1} H_{\text{gyro+act}}. \tag{3.95}$$

By expressing the yaw dynamics of the bare airframe as the frequency response H_{bare}, we can formulate the frequency response of the augmented yaw dynamics in terms of the frequency response of the different sub-elements, using the expression for the closed-loop system:

$$H_{\text{aug}} = \frac{H_{\text{bare}} H_{\text{act}}}{1 + H_{\text{bare}} H_{\text{act}} H_{\text{gyro}}}. \tag{3.96}$$

We can extract H_{bare} from this transfer function; since H_{aug}, H_{act}, and H_{gyro} are known, we can solve for the unknown bare airframe dynamics H_{bare} using frequency-response arithmetic

$$H_{\text{bare}} = \frac{H_{\text{aug}}}{H_{\text{act}} - H_{\text{aug}} H_{\text{act}} H_{\text{gyro}}}. \tag{3.97}$$

Figure 3.14 shows the magnitude of the frequency response of the augmented yaw dynamics, and the extracted bare airframe yaw dynamics. As can be seen from the figure, H_{bare} exhibits a pronounced resonance that does not agree with the standard first-order characteristic (3.94). This second-order characteristic probably results from a dynamic coupling between the bare airframe yaw dynamics and the engine drive-train dynamics. Therefore, for a detailed, and physically consistent, representation of the augmented yaw dynamics, the engine drive-train dynamics should be modeled in addition to the active yaw damping system. This would require measurement of the rotor speed and the engine throttle. Unfortunately we did not have enough time to undertake this work. An example of an identification model with engine drive-train dynamics is available in [23].

Augmented Yaw Dynamics Approximation

From Figure 3.14, we see that the augmented yaw dynamics has second-order characteristics with a small resonance. Thus, the augmented yaw dynamics can be accurately captured by a second-order system.

To avoid increasing the complexity of the model, we decided to revert to the representation we used in [60]. There, we assumed that the augmented yaw dynamics could be modeled as the first-order bare airframe dynamics with a yaw rate feedback represented by a simple first-order low-pass filter:

$$\frac{r_{fb}}{r} = \frac{K_r}{s + K_{r_{fb}}} \tag{3.98}$$

The closed-loop transfer function for the augmented yaw rate response becomes:

$$\frac{r}{\delta_{ped}} = \frac{N_{ped}(s + K_{r_{fb}})}{s^2 + (K_{r_{fb}} - N_r)s + (K_r N_{ped} - N_r K_{r_{fb}})} \tag{3.99}$$

with the corresponding differential equations used in the state-space model:

$$\dot{r} = N_v v N_r r + N_{ped}(\delta_{ped} - r_{fb}) \quad (3.100)$$

$$\dot{r}_{fb} = -K_{r_{fb}} r_{fb} + K_r r. \quad (3.101)$$

Again, since we have only the measurements of the pilot input δ_{ped} and the yaw rate r, the above representation is over-parameterized. One constraint between two parameters must be added in order to avoid identification problems due to correlated parameters. As a constraint, we stipulate that the pole of the low-pass filter must be twice as fast as the pole of the bare airframe yaw dynamics:

$$K_{r_{fb}} = -2N_r. \quad (3.102)$$

With this constraint, a low transfer function fitting cost is attained. However, because the constraint is based on an assumption, the value obtained for the bare airframe yaw damping derivative N_r is not necessarily physically meaningful. This should not constitute a big limitation, since the active yaw damping system can be retained in future flight-control designs as part of the bare airframe dynamics.

4.4 Heave Dynamics

The baseline stability derivative model for the heave dynamics (vertical motion) is:

$$\dot{w} = (-v_0 p + u_0 q) + Z_w w + Z_{col}\delta_{col}. \quad (3.103)$$

The terms in parentheses are the centrifugal forces that are present in cruise flight conditions. The control derivative Z_{col} accounts for the thrust perturbation due to collective blade pitch change δ_{col}. Note the speed derivative Z_w, which accounts for the rotor damping and fuselage drag. The resulting heave velocity to collective input transfer function is

$$\frac{w}{\delta_{col}} = \frac{Z_{col}}{s - Z_w}. \quad (3.104)$$

The rotor damping is the main dynamic effect that needs to be accounted for in the heave dynamics. This effect is present in all helicopters and results from the reduction of the aerodynamic angle of attack α caused by the increase in the inflow angle Φ following a increase in rotor thrust. A rotor that produces a steady thrust T in hover develops an inflow of velocity (from momentum theory [29])

$$v_{i,h} = \sqrt{\frac{T}{2\rho A}} \quad (3.105)$$

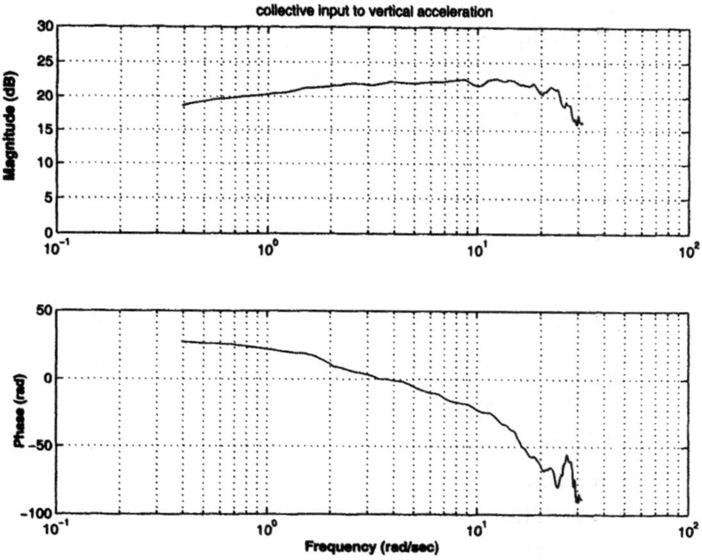

Figure 3.15. Frequency response of the vertical acceleration response to the collective control input $\dot{w}/\delta_{col}$.

where ρ is the air density, T is the rotor thrust, and A is the rotor disc area. The steady effect on the angle of attack is

$$\alpha = \theta - \Phi \approx \theta - \frac{1}{\Omega y}\sqrt{\frac{T}{2\rho A}} \tag{3.106}$$

which shows how the aerodynamic angle of attack is decreased with increasing thrust.

Figure 3.15 shows the estimated frequency response for the vertical acceleration $\dot{w}$ to the collective control input δ_{col}. Note that the magnitude is nearly flat. The damping of the heave dynamics is visible in the leading phase and the slightly rising magnitude with frequency (up to about 10 rad/sec). In particular, the response does not exhibit the peak in magnitude caused by the inflow and coning dynamics sometimes visible in full-size helicopters [23]. This confirms our earlier assumption that the coning dynamics could be neglected due to their fast time constant. This is a concequence of the faster rotor speed, and therefore higher natural frequency of the coning mode.

The inflow dynamics are also faster in a small rotor. A simple approximation to the time constant associated with the inflow dynamics is described in [68].

The approximation is derived by applying momentum theory while assuming that the mass of air being accelerated corresponds to 63.7 percent of the air contained in a sphere of the same diameter as the rotor. For axial flight the result for the inflow settling time is:

$$t_\lambda = \frac{0.849}{4\lambda_{i,trim}\Omega} \tag{3.107}$$

where $\lambda_{i,trim} = v_{i,trim}/\Omega R$ is the inflow ratio at the trim condition.

For the R-50 at hover, $\lambda_{i,h} = 0.045$ ($v_{i,h} = 20$ ft/sec), resulting in a settling time of $t_\lambda = 0.053$ sec. This value is about three times smaller than that obtained for a full-scale helicopter (UH-1H, $t_\lambda = 0.16$ sec). The corner frequency of the first-order approximation of the inflow dynamics is located at about $1/0.053 = 19$ rad/sec. This frequency is at the limit of the piloted excitations and significantly faster than the bandwidth of any heave controller.

4.5 Identification of the Actuator Dynamics

Swashplate actuators are used to implement the main rotor collective and cyclic control inputs for the rotor and stabilizer bar (cyclic only). To allow for an accurate identification of the vehicle, the dynamics of the swashplate actuators must be accounted for. If they are disregarded, their effects will be lumped into other parts of the vehicle model such as the rotor, and hence will bias the identification results. This is particularly important in our case, where the fast dynamics of the isolated rotor system are exposed by modeling the stabilizer bar.

The swashplate is moved by three identical actuators used for the implementation of the cyclic control and the collective control commands. The actuator dynamics were identified separately from the rest of the vehicle during ground experiments. A frequency sweep was applied as input to the actuator, and the deflection was measured by using encoders connected to the actuator output. The experiments were conducted unloaded, i.e. the actuator did not have to exert a force.

The frequency response of the actuator was first computed and subsequently used to estimate equivalent first- and second-order transfer functions. For the transfer function identification we used NAVFIT [88]. A time delay was also included as an identification parameter. The identification gave the following normalized transfer functions:

- the first-order transfer function is (time delay: $\tau_{act} = 0.0038\text{sec}$):

$$G_R(s) = \frac{15}{s + 15} \tag{3.108}$$

- the second-order transfer function is (time delay: $\tau_{act} = 0$ sec):

$$G_R(s) = \frac{39.18^2}{s^2 + 2 \cdot 1.22 \cdot 39.18 \cdot s + 39.18^2} \tag{3.109}$$

The second-order transfer function provides a marginally better fit than the first-order transfer function. Both transfer functions underestimate the magnitude in the mid-high frequencies and overestimate it in the high frequencies. They also overestimate the phase lag in the mid-frequencies. For the identification of the airframe dynamics, we used the simpler first-order transfer function.

In the future, it would be interesting to have access to the actuator deflection during the collection of flight data so that the bare airframe dynamics could be isolated from the actuators, and the effect of the load on the actuators could be accounted for.

5. Complete Parameterized Model

Now that all relevant vehicle dynamics have been modeled, the corresponding equations of motion can be combined to form the parameterized state-space model that will be used for the parametric identification.

5.1 Assembling the State-Space Model

The parameterized state-space model has the general form:

$$\boldsymbol{M}\dot{\boldsymbol{x}} = \boldsymbol{F}\boldsymbol{x} + \boldsymbol{G}\boldsymbol{u} \tag{3.110}$$

with state vector

$$\boldsymbol{x} = [u, v, p, q, \phi, \theta, a, b, w, r, r_{fb}, c, d]^T. \tag{3.111}$$

Note that the states of the dynamic augmentations (active yaw damping system and stabilizer bar) were placed last. The input vector is

$$\boldsymbol{u} = [\delta_{lat}, \delta_{lon}, \delta_{ped}, \delta_{col}]^T. \tag{3.112}$$

The system matrix $\boldsymbol{F}$ contains the stability derivatives, the input matrix $\boldsymbol{G}$ contains the input derivatives, and the $\boldsymbol{M}$ matrix contains the rotor and stabilizer bar time constants. The system can then be put in a standard state-space form

$$\dot{\boldsymbol{x}} = \boldsymbol{Ax} + \boldsymbol{Bu} \tag{3.113}$$

$$\boldsymbol{y} = \boldsymbol{Cx} \tag{3.114}$$

where $\boldsymbol{A} = \boldsymbol{M}^{-1}\boldsymbol{F}$, $\boldsymbol{B} = \boldsymbol{M}^{-1}\boldsymbol{G}$, and $\boldsymbol{C} = \boldsymbol{M}^{-1}\boldsymbol{H}$.

The entire vehicle dynamics are described by the following eleven coupled first-order differential equations:

- coupled fuselage-rotor-stabilizer longitudinal-lateral dynamics:
  - fuselage lateral and longitudinal motion $(\dot{v}, \dot{u})$ (Eqs. 3.79-3.80)
  - fuselage roll and pitch motion $(\dot{p}, \dot{q})$ (Eqs. 3.82-3.83)
  - rotor lateral and longitudinal flapping $(\dot{b}, \dot{a})$ (Eqs. 3.91-3.90)
  - stabilizer bar lateral and longitudinal flapping $(\dot{d}, \dot{c})$ (Eqs. 3.87-3.88)
- heave dynamics:
  - fuselage vertical motion $(\dot{w})$, (Eq. 3.81)
- augmented yaw dynamics:
  - fuselage yaw motion $(\dot{r})$ (Eq. 3.100)
  - yaw rate gyro feedback $(\dot{r}_{fb})$ (Eq. 3.101)

The resulting state-space model is illustrated in Figure 3.16. Two states were added (ϕ, θ) for, respectively, the roll and pitch Euler angles, resulting in a 13th-order system of equations. This particular model structure and parameterization corresponds to the one obtained after the model was already used for the identification of the Yamaha R-50. Therefore, it went through several iterations and refinements. The actuator dynamics are not included in the state-space model (although they could easily be incorporated). For the identification, the actuator dynamics were appended to the state-space model in form of already identified transfer functions.

5.2 Hover vs. Cruise Flight

A helicopter's response is different in hover flight than it is in cruise flight. In hover flight, the primary pilot task is to control the helicopter's attitude to produce the desired longitudinal and lateral forces that are required to position the helicopter in the three x, y, z dimensions. The helicopter's position, height above ground, and heading can be controlled independently. In cruise flight, the primary pilot task is to maintain the helicopter on a flight-path.

$$
\begin{bmatrix} \dot{u} \\ \dot{v} \\ \dot{p} \\ \dot{q} \\ \dot{\phi} \\ \dot{\theta} \\ \tau_f \dot{a} \\ \tau_f \dot{b} \\ \dot{w} \\ \dot{r} \\ \dot{r}_{fb} \\ \tau_s \dot{c} \\ \tau_s \dot{d} \end{bmatrix}
=
\begin{bmatrix}
X_u & 0 & 0 & 0 & 0 & -g & X_a & 0 & 0 & 0 & 0 & 0 & 0 \\
0 & Y_v & 0 & 0 & g & 0 & 0 & Y_b & 0 & 0 & 0 & 0 & 0 \\
L_u & L_v & 0 & 0 & 0 & 0 & 0 & L_b & L_w & 0 & 0 & 0 & 0 \\
M_u & M_v & 0 & 0 & 0 & 0 & M_a & 0 & M_w & 0 & 0 & 0 & 0 \\
0 & 0 & 1 & 0 & 0 & 0 & 0 & 0 & 0 & 0 & 0 & 0 & 0 \\
0 & 0 & 0 & 1 & 0 & 0 & 0 & 0 & 0 & 0 & 0 & 0 & 0 \\
0 & 0 & 0 & -\tau_f & 0 & 0 & -1 & A_b & 0 & 0 & 0 & A_c & 0 \\
0 & 0 & -\tau_f & 0 & 0 & 0 & B_a & -1 & 0 & 0 & 0 & 0 & B_d \\
0 & 0 & 0 & 0 & 0 & 0 & Z_a & Z_b & Z_w & Z_r & 0 & 0 & 0 \\
0 & N_v & N_p & 0 & 0 & 0 & 0 & 0 & N_w & N_r & N_{rfb} & 0 & 0 \\
0 & 0 & 0 & 0 & 0 & 0 & 0 & 0 & 0 & K_r & K_{rfb} & 0 & 0 \\
0 & 0 & 0 & -\tau_s & 0 & 0 & 0 & 0 & 0 & 0 & 0 & -1 & 0 \\
0 & 0 & -\tau_s & 0 & 0 & 0 & 0 & 0 & 0 & 0 & 0 & 0 & -1
\end{bmatrix}
\begin{bmatrix} u \\ v \\ p \\ q \\ \phi \\ \theta \\ a \\ b \\ w \\ r \\ r_{fb} \\ c \\ d \end{bmatrix}
+
\begin{bmatrix}
0 & 0 & 0 & 0 \\
0 & 0 & Y_{ped} & 0 \\
0 & 0 & 0 & 0 \\
0 & 0 & 0 & M_{col} \\
0 & 0 & 0 & 0 \\
0 & 0 & 0 & 0 \\
A_{lat} & A_{lon} & 0 & 0 \\
B_{lat} & B_{lon} & 0 & 0 \\
0 & 0 & 0 & Z_{col} \\
0 & 0 & N_{ped} & N_{col} \\
0 & 0 & 0 & 0 \\
0 & C_{lon} & 0 & 0 \\
D_{lat} & 0 & 0 & 0
\end{bmatrix}
\begin{bmatrix} \delta_{lat} \\ \delta_{lon} \\ \delta_{ped} \\ \delta_{col} \end{bmatrix}
$$

Figure 3.16. **Parameterized State-Space Model**

The core part of the helicopter dynamics, the attitude dynamics, do not change significantly from the hover flight conditions. However, the helicopter's responses following attitude changes is dramatically different in cruise flight; becoming similar to a fixed-wing aircraft. For example, the pitch angle affects the vertical acceleration, giving the cyclic longitudinal the role of an airplane's elevator; the roll angle affects the lateral acceleration as well as the heading, giving to the cyclic lateral the role of ailerons.

Fortunately, no major modifications to the parameterized model are necessary to accomodate the cruise flight condition. We added only the following derivatives:

- M_{col} to account for the elevator-like effect of the collective control on the helicopter pitch.
- Y_{ped} to account for the rudder-like effect of the pedal control on the helicopter lateral acceleration (lateral-directional effects).
- M_w and L_w to account for the effects of the vertical speed on the pitch and roll dynamics.

We also dropped some of the hover derivatives (see Table 4.3 showing the identified derivatives for details).

Moreover, in cruise flight the nonlinear centrifugal acceleration terms that appear in the equation of motion of the fuselage accelerations (Eqs. 3.79 to 3.81) need to be accounted for. We obtained good approximation for the cruise flight trim conditions by averaging the cruise speeds over all cruise flight experiments. This results in the following additional derivatives (values for the R-50):

$$\begin{aligned} X_q = -Y_p = & \quad -w_0 & = 0 \text{ ft/sec} \\ X_r = -Z_p = & \quad v_0 & = -11 \text{ ft/sec} \\ Y_r = -Z_q = & \quad -u_0 & = -49 \text{ ft/sec.} \end{aligned}$$

Chapter 4

IDENTIFICATION OF THE MODEL

Now that we have estimated the frequency responses and developed a linear parameterized model for our small-scale rotorcraft, we can proceed with the identification of the model parameters. In this chapter, we summarize the identification process and present the results obtained for the Yamaha R-50. We first explain how the estimated frequency responses are related to the transfer functions derived from the state-space model, and also describe how to account for the kinematic effects introduced by a sensor position offset. Then, we describe the identification procedure we developed for the R-50, i.e. how we broke down the identification of the state-space model into subsystems, selected the frequency responses and frequency ranges, and refined the model parameterization. Finally, we present the frequency response fit, discuss the identified model parameters, demonstrate the accuracy of the model through a time-domain validation, and, to conclude, provide a comparison of the key identified parameters' values with the values predicted by helicopter theory.

1. Identification Setup

The parameters of the state-space model are identified through a fitting process between the estimated frequency responses and the transfer functions that are calculated from the state-space model of Eq. 3.110 using Eq. 2.27. In the following, we describe how to correctly relate the measured data with the outputs of the state-space equations and describe potential issues with flight-data kinematics and how to address them.

1.1 Setup of the Output Equations

The flight-data measurements consist of the raw measurements from the inertial measurement system accelerometers and angular rate gyroscopes. When the accelerometers are placed at the vehicle center of gravity, the sensors measure the accelerations the vehicle experiences as a result of the various forces acting on it. The corresponding mathematical expression is

$$\boldsymbol{a}_{\text{meas}} = \boldsymbol{a}_{\text{cg}} = \frac{d^I m\boldsymbol{v}}{dt} = \begin{bmatrix} a_x \\ a_y \\ a_z \end{bmatrix}_{\text{meas}} = \frac{\boldsymbol{F}_{\text{aero}}}{m}. \tag{4.1}$$

The inertial accelerations expressed in terms of the time rate of change of the body velocities are

$$\frac{d^I m\boldsymbol{v}}{dt} = \frac{d^B m\boldsymbol{v}}{dt} + \boldsymbol{\omega} \times \boldsymbol{v} = \frac{\boldsymbol{F}_{\text{ext}}}{m} = \frac{\boldsymbol{F}_{\text{grav}}}{m} + \frac{\boldsymbol{F}_{\text{aero}}}{m} \tag{4.2}$$

where $\boldsymbol{F}_{\text{ext}}$ are the total external forces acting on the vehicle and are made of the sum of the aerodynamic $\boldsymbol{F}_{\text{aero}}$ and gravitational forces $\boldsymbol{F}_{\text{grav}}$.

Using these two equations, we can relate the measured accelerations and the time rate of change of the vehicle body velocities

$$\boldsymbol{a}_{\text{meas}} = \begin{bmatrix} a_x \\ a_y \\ a_z \end{bmatrix}_{\text{meas}} = \dot{\boldsymbol{v}} + \boldsymbol{\omega} \times \boldsymbol{v} - \frac{\boldsymbol{F}_{\text{grav}}}{m}. \tag{4.3}$$

We see that when fitting the frequency responses estimated from the acceleration measurements with the transfer function derived from the state-space model (which describes the time rate of change of the body velocities $\dot{\boldsymbol{v}}$), we must set up output equations to account for the centripetal effects $(\boldsymbol{\omega} \times \boldsymbol{v})$ and the gravitational forces.

The angular rate gyroscopes sense the vehicle's angular velocities $\boldsymbol{\omega}_{\text{meas}} = [p, q, r]^T$. Therefore, the measurements, and thus the estimated frequency responses, for the angular dynamics are consistent with the attitude equations of motion.

1.2 Motion Sensor Kinematics

In Eq. 4.3, we assume that the acceleration measurements are the accelerations of the vehicle's center of gravity (c.g.) $\boldsymbol{a}_{\text{cg}}$. However, the IMU that houses the three accelerometers is often offset from the the vehicle center of gravity as shown in Fig. 4.1. Therefore, the accelerations sensed are those of

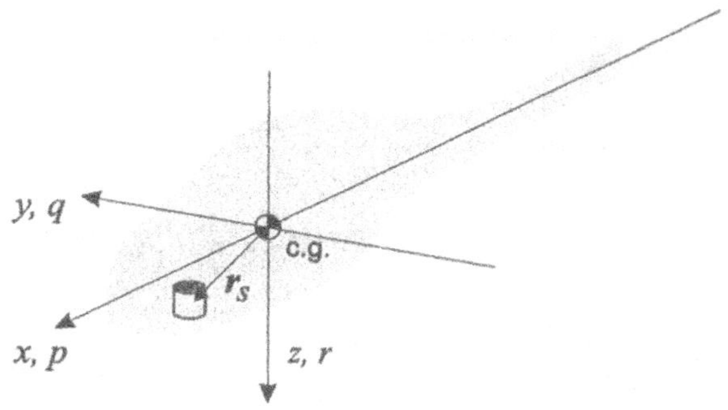

Figure 4.1. **Illustration of the offset between the sensor position (IMU) and the vehicle center of gravity. The position of the sensor is given by the position vector $\boldsymbol{r}_s$**

the attachement point on the helicopter fuselage. From the kinematics of relative motion we know the acceleration at a point S (position vector $\boldsymbol{r}_S$) fixed on a rigid object moving relative to an inertial reference frame is given by:

$$\boldsymbol{a}_{\text{meas}} = \boldsymbol{a}_{\text{cg}} + \boldsymbol{\omega} \times (\boldsymbol{\omega} \times \boldsymbol{r}_S) + \dot{\boldsymbol{\omega}} \times \boldsymbol{r}_S. \tag{4.4}$$

We see that for a non-zero offset $\boldsymbol{r}_S$, the measured accelerations are biased by a centripetal $\boldsymbol{a}_{\text{cent}} = \boldsymbol{\omega} \times (\boldsymbol{\omega} \times \boldsymbol{r}_S)$ and a tangential acceleration component $\boldsymbol{a}_{\text{tang}} = \dot{\boldsymbol{\omega}} \times \boldsymbol{r}_S$.

When the location of the IMU with respect to the c.g. is known, the acceleration measurements can be corrected for the effect of the offset prior to the estimation of the frequency responses. This is simply done by subtracting the two biases $\boldsymbol{a}_{\text{cent}}$ and $\boldsymbol{a}_{\text{tang}}$ from $\boldsymbol{a}_{\text{meas}}$. Figure 4.2 shows the reconstructed tangential and centripetal acceleration biases due to a sensor offset in MIT's X-Cell .60 helicopter (the measured offset is $\boldsymbol{r}_S = [0.1, 0, 0.125]^T$ m). The measurements are from a lateral sweep experiment, which induces strong roll motions. Notice that the tangential bias in the body y-direction has the largest magnitude for this experiment. Figure 4.3, shows the cyclic input applied during the sweep and compares the raw measurement data with the corrected data.

When the position of the center of gravity is only known approximately, a different approach must be taken. One possibility is to estimate the offset and other parameters of the sensor system (e.g. sensor biases), before using the measurements for system identification, using optimal estimation methods. The other possibility is to add kinematic equations to the identification model and include the offset as an identification parameter. Or, if the data was corrected for an offset but the c.g. location was only know approximately, we

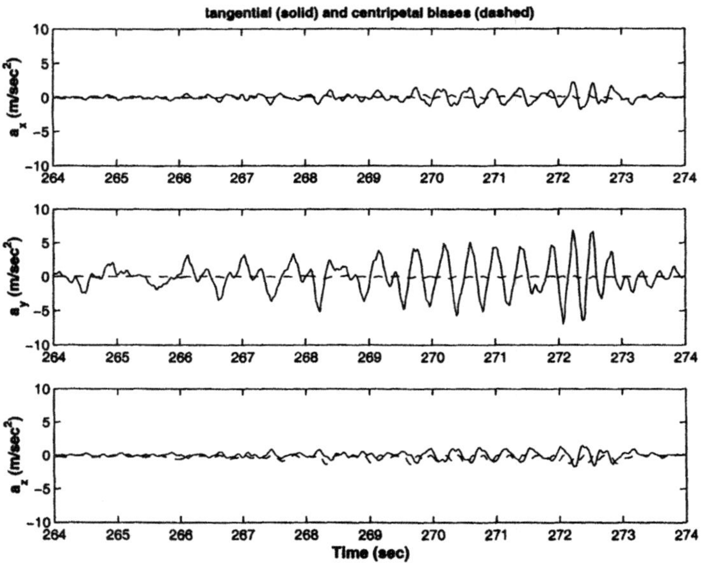

Figure 4.2. Reconstructed tangential (solid) and centripetal (dashed) acceleration biases due to sensor offset.

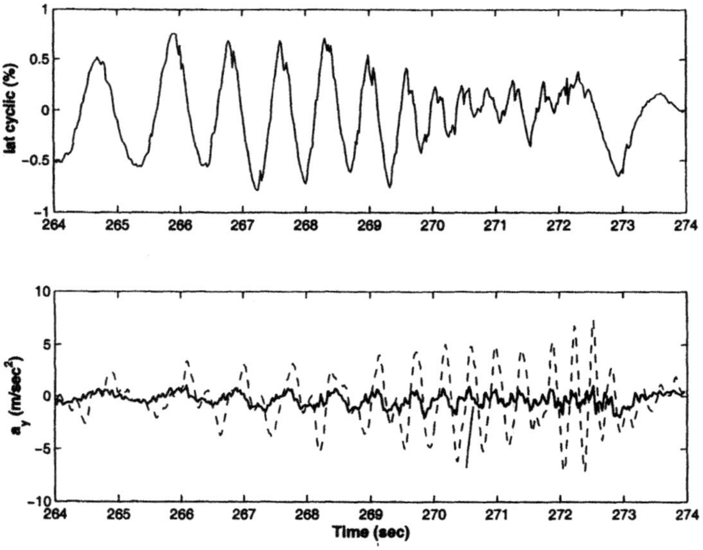

Figure 4.3. Effect of sensor offset on acceleration measurement during a lateral cyclic sweep. The top plot shows the cyclic input applied during the experiment, and the bottom plot shows the raw measurement data (dashed) and the corrected measurement data (solid).

can add kinematic equations for the effect in an error in the c.g. position. This latter approach was used for Carnegie Mellon's Yamaha R-50, where the data was already corrected for a sensor offset, but the vertical c.g. location was only known up to an uncertainty h_{cg}. In the following, we first illustrate how an offset can corrupt the identified derivatives, and then we introduce the kinematic correction term used to explicitly take sensor offsets into account.

1.3 Effects of Flight-Data Kinematics

The R-50 is balanced so that its center of gravity is below the rotor hub, hence the remaining offset that needs to be accounted for is due to the uncertainty in the vertical c.g. location. We can write the residual offset as $\boldsymbol{r}_S = [0, 0, h_{\text{cg}}]^T$. In the following, we illustrate the effect that biases due to such a vertical offset can have on the identified derivatives.

For the analysis, we dismiss the centripetal bias, which is usually significantly smaller than the tangential bias. The tangential acceleration bias in this case is

$$\boldsymbol{a}_{\text{tang}} = \dot{\boldsymbol{\omega}} \times \boldsymbol{r}_S = \begin{bmatrix} \dot{p} \\ \dot{q} \\ \dot{r} \end{bmatrix} \times \begin{bmatrix} 0 \\ 0 \\ h_{\text{cg}} \end{bmatrix} = \begin{bmatrix} \dot{q}h_{\text{cg}} \\ -\dot{p}h_{\text{cg}} \\ 0 \end{bmatrix}. \tag{4.5}$$

To account for the fact that the acceleration measurements are biased, we write the longitudinal and lateral equations of motion Eqs. 3.79 and 3.80 in terms of the measured velocity rates and velocities

$$\dot{u}_{\text{meas}} = X_u u_{\text{meas}} - g\theta + X_a a \tag{4.6}$$

$$\dot{v}_{\text{meas}} = Y_v v_{\text{meas}} + g\phi + Y_b b. \tag{4.7}$$

We can define the measured time rate of change of the x- and y- body velocities using the definition for the tangential bias from Eq. 4.5

$$\dot{u}_{\text{meas}} = \dot{u} + h_{cg}\dot{q} \tag{4.8}$$

$$\dot{v}_{\text{meas}} = \dot{v} - h_{cg}\dot{p}. \tag{4.9}$$

Similarly, the measured velocities are

$$u_{\text{meas}} = u + h_{cg}q \tag{4.10}$$

$$v_{\text{meas}} = v - h_{cg}p. \tag{4.11}$$

The effect of the tangential bias on the identification can be illustrated by replacing the velocities and velocity rates with the expressions for the biased velocities and velocity rates

$$\dot{u} + h_{cg}\dot{q} = X_u(u + h_{cg}q) - g\theta + X_a a \quad (4.12)$$

$$\dot{v} + h_{cg}\dot{p} = Y_v(v - h_{cg}p) + g\phi + Y_b b. \quad (4.13)$$

The fuselage angular rates can be replaced by the expressions from Eqs. 3.82 and 3.83. After regrouping the terms, we obtain the following equations written in terms of the true c.g. velocities and velocity rates

$$\dot{u} = X_u h_{cg} q - g\theta + (X_u - h_{cg}M_u)u + (X_a - h_{cg}M_a)a \quad (4.14)$$

$$\dot{v} = Y_v h_{cg} p - g\phi + (Y_v - h_{cg}L_v)v + (Y_b - h_{cg}L_b)b. \quad (4.15)$$

We can see that both the speed derivatives X_u and Y_v and the rotor spring derivatives X_a and Y_b are biased. For example, based on physical principles, the X_a and Y_b derivatives should theoretically be equal, respectively, to plus and minus the value of the gravity ($g = 32.2$ ft/s^2). Instead, with the bias, we would approximately obtain

$$\tilde{X}_a = X_a - h_{cg}M_a \quad (4.16)$$

$$\tilde{Y}_b = Y_b + h_{cg}L_b. \quad (4.17)$$

The two rotor derivatives are good indicators of problems with the acceleration measurements. Moreover, if the data have been properly corrected, we can introduce the constraints $X_a = -g$ and $Y_b = g$, which reduces the number of derivatives to identify.

For the R-50, we explicitly accounted for a vertical offset h_{cg} in the identification model by augmenting the model with the bias equations Eqs 4.8 and 4.9 and writing the longitudinal and lateral equations of motion in terms of the measured velocity rates. Using this approach, we were able, at the same time, to constrain the rotor derivatives and identify the value for the unknown offset h_{cg}.

2. Identification Process

The entire identification process is performed using the state-space identification tool (DERIVID) in CIFER [87]. We broke down the identification of the entire state-space model into intermediate steps. The intermediate steps helped us obtain preliminary estimates for key derivatives (e.g. rotor time constant, flapping spring derivatives) and allowed an incremental verification and refinement of the parameterized model. The parameterized model presented in the previous chapter (Eq. 3.16) corresponds to the final form that was obtained after successful completion of the entire identification process. The identification process also involves selecting the frequency responses and frequency ranges used for the model fitting and deciding which parameters need to be constrained. For the verification and refinement, we used statistical tools to determine if some derivatives were insensitive, difficult to identify due to correlation with other derivatives, or had high standard deviations. The process was repeated until satisfactory results were attained. In the following, we first describe the selection of the frequency responses and frequency ranges, then we explain how the identification was broken down and describe the statistical tools used for the model refinement.

2.1 Selection of the Frequency Responses

To determine which frequency responses and frequency range to use for the identification, we relied primarily on the coherence metric that was calculated during the multivariable frequency domain analysis. We included all frequency responses that had a coherence γ^2 of 0.6 or higher, and we selected the frequency range for the fitting of each response based on that same coherence threshold. Table 4.1 summarizes which responses and respective frequency ranges were selected for the identification of both the hover and cruise-flight condition. The fitting range was in some cases adjusted later based on the fit achieved during the identification. The frequency range selected for the identification should of course also satisfy the accuracy requirements.

2.2 Breakdown of the Identification

By breaking down the identification into a series of smaller identification problems, we were able to better determine if the model structure allowed a proper fit of the estimated frequency responses, i.e. if the model was able to capture the dynamics of the vehicle with a sufficient level of precision. For each step, a series of key derivatives were identified, and the focus was to obtain a good fit of the estimated frequency response and meaningful values for those derivatives. The emphasis is first put on the key on-axis responses shown in Table 4.1. In some cases we used the intermediate results to refine the model structure and adjust the fitting frequency range. The criteria and statistics used for the model refinement will be described in the next section. At each step the values of the derivatives identified during the prior step are used as

Table 4.1. **Frequency responses and frequency ranges (in rad/sec) selected for the identification process. (First row: hover condition; second row: cruise-flight condition).**

	δ_{lat}	δ_{lon}	δ_{col}	δ_{ped}
p	0.32 - 20	0.4 - 20	-	-
	0.7 - 20	0.7 - 7	2.4 - 16	-
q	0.4 - 20	0.32 - 20	-	-
	1.6 - 8	0.7 - 14	0.7 - 18	-
r	0.4 - 1.6	-	0.7 - 3	0.5 - 8
	0.7 - 11	-	-	2 - 8
a_x	0.4 - 8	0.32 - 14	-	-
		0.7 - 12	-	-
a_y	0.32 - 20	-	-	-
	0.7 - 10	-	-	-
a_z	3 - 16	6 - 16	0.32 - 30	4 - 8
	0.7 - 5	0.7 - 24	0.7 - 15	-
u	0.4 - 8	0.32 - 14	-	-
	-	-	-	-
v	0.32 - 16	0.4 - 16	-	-
	0.63 - 9	-	-	-
w	-	-	-	-
	1.4 - 9	0.63 - 12	-	-

initial values. Also additional dynamics and the off-axes dynamics are added to the identification problem. Eventualy the entire state-space model is identified.

The key derivatives are those that can be identified at the respective step and that will only experience small changes as the remaining dynamics are added to the problem. The identification steps and key derivatives are the following:

1 Decoupled roll and pitch dynamics without explicit stabilizer equations of motion
Key derivatives: effective rotor time constant (τ_{eff}), flapping spring derivatives (M_a, L_b), effective input derivatives ($A_{lon,\text{eff}}$, $B_{lat,\text{eff}}$)

2 Coupled roll-pitch equations
Key derivatives: cross-coupling derivatives (A_b,B_a), effective input derivatives ($A_{lat,\text{eff}}$, $B_{lon,\text{eff}}$)

3 Addition of stabilizer bar dynamics
Key derivatives: stabilizer bar time constant (τ_s) and main blade time constant (τ_f); Bell mixer gearing (A_c and B_d); cyclic input sensitivities ($A_{lat}, A_{lon}, B_{lat}, B_{lon}$ and D_{lat}, C_{lon})

4 Translational longitudinal and lateral equations of motion
Key derivatives: speed derivatives (X_u and Y_v) and sensor offset (h_{cg})

5 Heave dynamics
Key derivatives: heave damping (Z_w) and collective input sensitivity (Z_{col})

6 Yaw dynamics
Key derivatives: yaw damping (N_r), yaw rate feedback (K_r), and pedal input sensitivity (N_{ped})

7 Coupling of heave and yaw dynamics
Key derivatives: collective-to-yaw and yaw-to-collective derivatives (N_{col} and Z_r, respectively)

8 Entire state-space model; coupling of the longitudinal and lateral translational dynamics with the heave and yaw dynamics

While this procedure was developed for the Yamaha R-50, it has proven to be practical for the identification of the X-Cell helicopter. Once the frequency responses were identified, only the frequency fitting ranges had to be adjusted at each step.

2.3 Model Refinements

The final model structure is obtained by adding and removing derivatives according to: (i) level of frequency response agreement (frequency error costs), (ii) statistical metrics from the model parameters (insensitivity and Cramer-Rao percent), (iii) level of agreement with the system's time responses (time domain verification), and (iv) level of agreement of the parameter with its theoretical value (when a specific model parameter has a defined physical meaning).

The useful statistics are: the insensitivity of the cost function to each derivative; the Cramer-Rao (CR) bound, which gives a lower bound on the achievable standard deviation (indicating how accurately a parameter can be identified); and the correlation among the derivatives. Insensitive parameters do not affect the value of the cost function, i.e. they can take any value without it being reflected in the fitting cost, hence they should be dropped. Parameters with high

CR bounds cannot be identified accurately and hence if possible they should fixed at a known value or dropped from the parameterized model. With regard to correlated parameters, we need to drop or, when possible, fix those derivatives to a value determined theoretically or by means of other experiments. To precisely determine the correlation among the derivatives and help us with this task, we use a confidence ellipsoid

The statistics used for the model refinement, including the insensitivity, CR bound, correlation, and confidence ellipsoid, are derived from the Hessian matrix $\boldsymbol{H}$:

$$\boldsymbol{H} = \frac{\partial^2 J(\boldsymbol{\Theta})}{\partial \boldsymbol{\Theta}^2} \tag{4.18}$$

where $J(\boldsymbol{\Theta}) = \boldsymbol{\varepsilon}(\boldsymbol{\Theta})^T \boldsymbol{W} \boldsymbol{\varepsilon}(\boldsymbol{\Theta})$ is the cost function for the vector of unknown parameters $\boldsymbol{\Theta}$ (see Eq. 2.26).

3. Identification Results

In this section, we present a summary of the identification results for the Yamaha R-50. The three components of the identification results are: the frequency response agreement, i.e. fit between the estimated frequency responses and the model predicted responses; the identified model parameters; and, finally, the level of accuracy of the time response predictions.

3.1 Frequency Response Agreement

The frequency responses predicted by the identified model show good agreement with the frequency responses derived from the flight data. This is the case for both the hover and the cruise-flight conditions. The transfer function costs attained for the different input-output pairs are given in Table 4.2, and the comparison between the frequency responses is provided in Figure 4.4 for the hover-flight condition and Figure 4.5 for the cruise-flight condition. Compared with the results that were obtained for the first attempt at modeling the R-50, where the rotor-stabilizer system was modeled as a simple rotor [60], the off-axis angular responses have been significantly improved (the transfer function cost dropped about 50% for q to δ_{lat} and p to δ_{lon}). This clearly demonstrate the value of including the stabilizer bar in the modeling (see Chapter 5 for more details).

The close agreement achieved for the small-scale Yamaha R-50 helicopter (average cost of 31.5 and 33.9 for the hover and cruise conditions, respectively) is better than what is usually achieved for full-scale helicopters (for full-scale

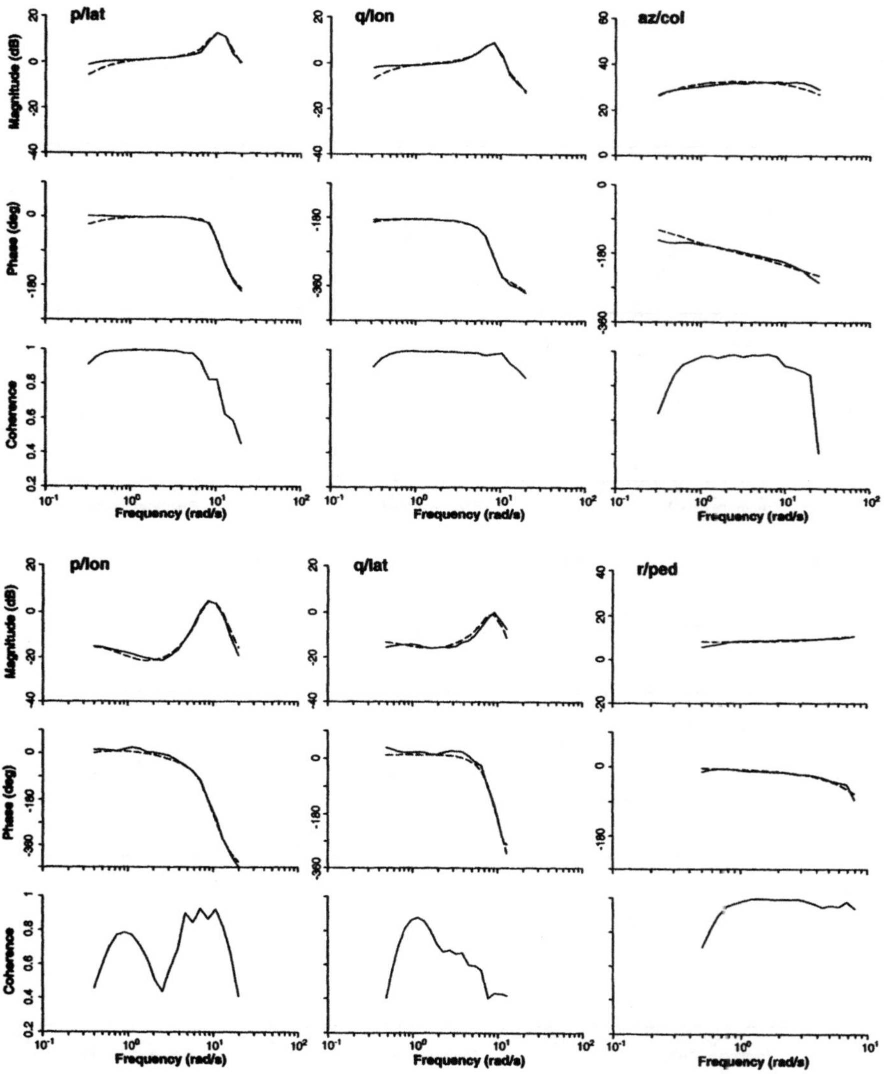

Figure 4.4. **Comparison between the frequency responses computed from the flight data (in solid) and those predicted by the identified model (in dashed); the results are for the hover-flight condition.**

rotorcraft, a frequency response cost of 70 is considered to be an excellent fit [89]). In particular, it is important to note that the good results are attained with a relatively simple model. This indicates that the dynamics of small-scale helicopters could be subject to fewer higher-order effects than are found in full-

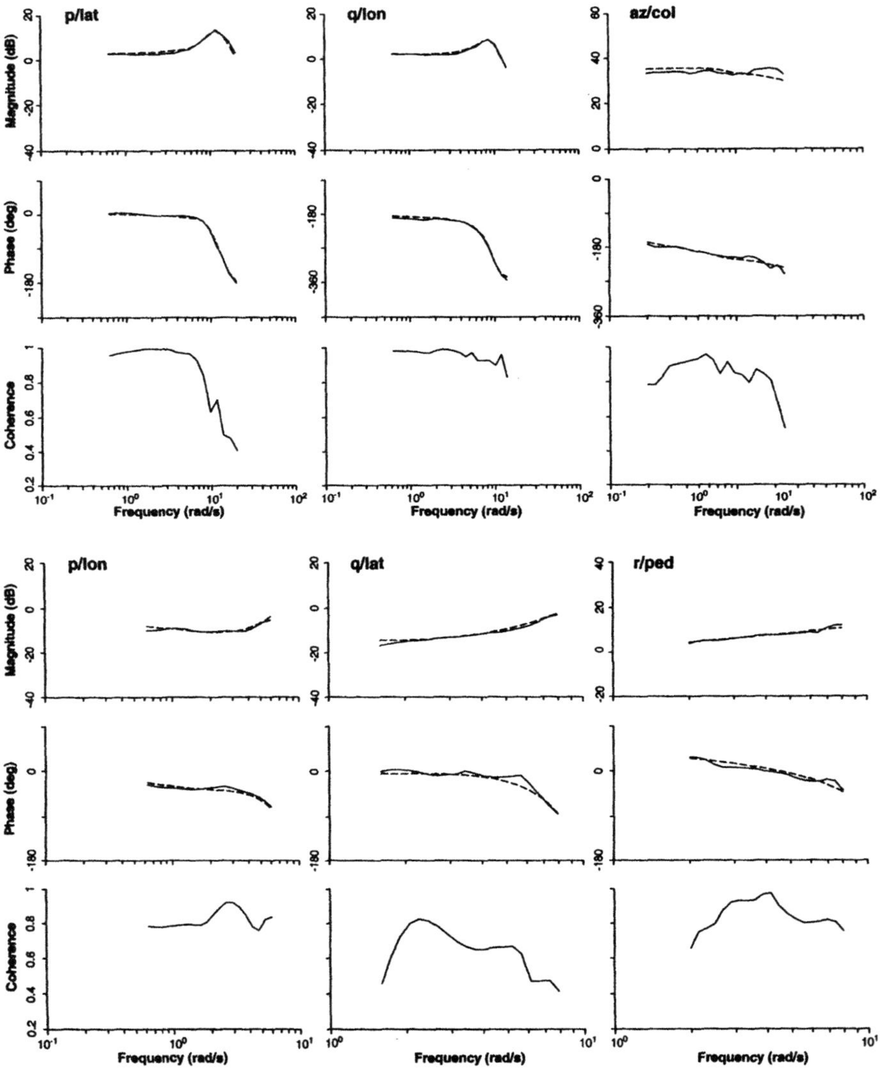

Figure 4.5. Comparison between the frequency responses computed from the flight data (in solid) and those predicted by the identified model (in dashed); the results are for the cruise-flight condition ($u_0 = 49$, $v_0 = -11$, and $w_0 = 0$ ft/sec).

scale rotorcraft (e.g. fuselage aerodynamic effects, high-order rotor dynamics). For example, the R-50's attitude dynamics have a distinct second-order effective nature (see p to δ_{lon} and q to δ_{lat} in Figure 4.4), which can be accurately fitted

Table 4.2. Transfer function costs attained for each input-output pair during the identification.

	Hover	Cruise		Hover	Cruise
vx/lat	15.7	-	q/lon	41.3	9.7
vy/lat	18.4	-	ax/lon	27.9	48.8
vz/lat	-	71.1	ay/lon	35	-
p/lat	61.5	13	az/lon	33.8	44.2
q/lat	50.4	18.5	p/col	-	45.8
ax/lat	15.7	-	q/col	-	19.1
ay/lat	17.4	23.5	r/col	35.2	-
r/lat	27.4	-	az/col	55.3	78.2
az/lat	25	-	ay/ped	-	10.3
vx/lon	27.9	-	vy/ped	-	39.1
vy/lon	35	-	r/ped	18.8	23.5
vz/lon	-	48.5	az/ped	19.4	-
p/lon	37.1	15.6	**Average**	**31.5**	**33.9**

by the second-order fuselage-rotor hybrid model, or the third-order fuselage-rotor-stabilizer bar model for an even better fit and more meaningful result.

3.2 Identified Model Parameters

The numerical values for the identified derivatives for both the hover and the cruise-flight conditions are presented in Table 4.3. The parameters that were constrained are presented in Table 4.4. Each derivative value is shown together with its CR bound and insensitivity (both expressed in percentages). Overall, the statistics indicate that all of the key control and response derivatives were extracted with a high degree of precision. In two cases, the quasi-steady derivatives (see X_u and N_v) and the main rotor time constant (τ_f) came out with a CR percent that exceeded the guidelines of 30%. For each identified derivative, the Table also provides the ratio between the value identified in forward flight and the value identified in hover flight. This gives a measure of how much the derivatives change between the two flight conditions. In the following, we discuss these results qualitatively. Later, in Section 4, we will provide a theoretical validation of the key derivatives.

Rotor and Attitude Dynamics Derivatives

For the hover condition, the identified stabilizer bar and bare rotor time constants came out as, respectively, $\tau_s = 0.34$ sec and $\tau_f = 0.046$ sec. For cruise

Table 4.3. Identified model parameters for the hover- and cruise-flight condition ($u_0 = 49$, $v_0 = -11$, and $w_0 = 0$ ft/sec).

	Hover Flight			Cruise Flight			ratio
	Value	CR %	Insens. %	Value	CR %	Insens. %	
M-Matrix							
τ_f	0.046	10.9	1.64	0.0346	31.7	2.37	0.75
h_{cg}	-0.411	6.25	1.78	-0.321	14.7	6.695	0.78
τ_s	0.342	7.35	0.828	0.259	6.52	0.786	0.76
F-Matrix							
X_u	-0.0505	62.3	29.6	-0.122	57.7	27.5	2.41
Y_v	-0.154	22.9	10.9	-0.155	32.7	6.84	1.00
L_u	-0.144	12.3	3.07	-	-	-	-
L_v	0.143	19.5	6.54	-	-	-	-
L_w	-	-	-	-0.213	15.3	3.96	-
L_b	166	1.87	0.6	213	0.00145	2.14	1.28
M_u	-0.0561	20.2	5.63	-	-	-	-
M_v	-0.0585	14.5	4.23	-	-	-	-
M_w	-	-	-	0.0728	21.2	5.68	-
M_a	82.6	6.28	0.592	108	0.0593	0.786	1.31
B_a	0.368	10.5	1.13	0.419	11.5	2.18	1.14
B_d	0.71	4.11	0.782	0.664	9.66	1.55	0.93
A_b	-0.189	11.7	4.47	-0.176	21.9	9.39	0.93
A_c	0.644	9.49	0.819	0.577	7.73	1.09	0.89
Z_b	-131	2.77	1.62	-	-	-	-
Z_a	-9.75	19.9	8.26	-	-	-	-
Z_w	-0.614	10.5	4.47	-1.01	4.72	2.07	1.65
Z_r	0.93	8.15	2.75	-	-	-	-
N_p	-3.53	14.2	3.66	-	-	-	-
N_v	0.0301	32.5	9.08	0.401	8.8	3.36	13.3
N_w	0.0857	14.1	5.09	-	-	-	-
N_r	-4.13	9.71	2.79	-3.9	10.6	3.67	0.94
K_r	2.16	4.42	1.74	2.18	7.75	2.7	1.00
G-Matrix							
B_{lat}	0.14	7.06	1.61	0.124	16.3	2.64	0.88
B_{lon}	0.0138	13.4	4.21	0.02	17.8	6.7	1.45
A_{lat}	0.0313	7.89	2.06	0.0265	8.19	2.54	0.85
A_{lon}	-0.1	9.21	1.18	-0.0837	13.9	1.87	0.83
Z_{col}	-45.8	4.32	1.84	-60.3	4.06	1.86	1.31
M_{col}	-	-	-	6.98	6	1.43	-
N_{col}	-3.33	10.6	3.67	-	-	-	-
N_{ped}	33.1	5.45	1.92	26.4	7.18	2.44	0.80
D_{lat}	0.273	12.2	1.89	0.29	12.5	2.36	1.06
C_{lon}	-0.259	10.7	1.62	-0.225	11.2	1.66	0.87
Y_{ped}	-	-	-	11.23	21.9	4.66	-
τ_{ped}	0.0991	13.2	6.09	0.0589	21.8	6.93	0.59

Table 4.4. **Constrained parameters for hover- and cruise-flight conditions.**

	Hover Flight		Cruise Flight	
	Value	Constraint	Value	Constraint
$N_{r_{fb}}$	-33.1	$-N_{ped}$	-26.4	$-N_{ped}$
$K_{r_{fb}}$	-8.26	$2N_r$	-7.79	$2N_r$
X_θ, X_a	-32.2	$-g$	-32.2	$-g$
X_r	-	-	-11.0	v_0
Y_ϕ, Y_b	32.2	g	32.2	g
Y_r	-	-	-49.2	$-u_0$
Z_p	-	-	11.0	$-v_0$
Z_q	-	-	49.2	u_0

flight, the rotor time constants decreased to $\tau_s = 0.26$ sec and $\tau_f = 0.035$ sec. The smaller time constants in cruise flight could be a consequence of changes in the rotor inflow [36] (see Eq. 4.22); however other effects could also be involved.

The identified roll and pitch rotor spring derivatives are, respectively, $L_b = 166\ 1/\text{sec}^2$ and $M_a = 82.6\ 1/\text{sec}^2$ for the hover-flight condition. Their values for cruise-flight are about 30% larger: $L_b = 213\ 1/\text{sec}^2$ and $M_a = 108\ 1/\text{sec}^2$. The higher control authority in cruise flight could, in part, be due to the higher rotor thrust setting necessary to balance the aerodynamic fuselage and rotor drag. Moreover, the cruise-flight tests took place at lower ambient air temperature, which could also have influenced the results (changing the stiffness of the elastomeric blade restraints and thus affecting the rotor spring derivatives).

The lateral and longitudinal cyclic control derivatives have reasonable values. Their values do not change significantly with the flight conditions except for B_{lon}, which has a 45% larger value in cruise-flight. This increase indicates a higher cross-axis influence of the longitudinal input on the roll dynamics (changes in the rotor coning angle following changes in load factor result in a roll motion; transverse flow effect [29]). It is difficult to come up with a good explanation for these changes without a detailed treatment of the dynamics in forward flight.

All other cyclic control derivatives change only moderately (between 6 and 17%). The smallest changes occur for the stabilizer cyclic control derivatives (between 6 and 13%).

For the cruise-flight condition, a collective-to-pitch control derivative ($M_{col} = 6.98\ \text{rad/sec}^2\text{-\%}$) was added. This derivative accounts for the pitch moment generated by the application of collective pitch. This effect is due to the greater lift on the advancing blade than on the retreating blade in forward flight [68].

Speed Derivatives

The speed force damping derivatives X_u, Y_v in the linear motion equations (Eq. 3.79) and (Eq. 3.80) have the sign and relative magnitudes expected for hovering helicopters. An interpretation of their identified values is not meaningful because of their high insensitivity (between 11% and 30%) and very high CR percent (between 23% and 62%). Therefore, the identified values are unreliable. This weak result is due to the insufficient low-frequency information content of the flight data. A better identification of these derivatives could be achieved by increasing the record lengths. This, however, is not feasible for piloted experiments.

The speed moment derivatives L_u, L_v and M_u, M_v in the angular rate Eqs. 3.82 and 3.83 are important indicators for the speed-stability characteristics. Note that for hover flight they have been extracted with a higher degree of precision than have the force damping derivatives. However, these derivatives were dropped for the cruise condition because of high insensitivity. L_v and M_u describe how the helicopter's attitude responds to an increase in airspeed. Helicopters normally have the tendency to roll or pitch away from the direction of the speed perturbation (caused by an increase in vehicle speed and/or a wind gust). This effect, referred to as dihedral effect, is due to the lift asymmetry between the advancing and the retreating blade that occurs when the rotor is exposed to a velocity perturbation, as well as to the 90-degrees azimuth shift of the rotor flapping response. For example, when a hovering helicopter experiences a wind gust from the right side (y axis, relative airspeed $v > 0$), it will experience a rolling motion to the left (roll rate $p < 0$). The values that were identified, $L_v = 0.14$ rad/sec-ft and $M_u = -0.056$ rad/sec-ft, both have their sign opposite to what is normally expected – meaning they have a destabilizing effect on the speed. Their magnitude, however, is small (for 49 ft/sec, the cruise condition longitudinal speed, the effect of the longitudinal speed moment derivative is equivalent to less than 2 degrees of longitudinal flapping). The explanation for this could be the presence of the stabilizer bar, which, besides damping the roll and pitch motion, also reduces the sensitivity of the rotorcraft to airspeed perturbations and reduces the natural frequency of the phugoid mode (in the present case it results in a long period unstable phugoid mode as we will see later in Chapt. 5). This also means that less longitudinal pilot input is necessary to trim the rotorcraft in cruise flight.

Directional stability of the helicopter is described by the directional stability derivative N_v. The tail rotor and tail fin give the helicopter a weathervane behavior (tendency of the helicopter to turn its tail away from the direction of the speed perturbation). The values that were identified – $N_v = 0.03$ rad/sec–ft for hover and $N_v = 0.4$ rad/sec–ft for cruise – have the correct sign, and the increase for the cruise-flight condition is expected due to the higher dynamic pressure.

With the help of the offset equations (Eq. 4.8) and (Eq. 4.9), we were able to constrain the force coupling derivatives to gravity ($-X_a = Y_b = g$) and, at the same time, identify the vertical c.g. offset, which came out to be $h_{cg} = -0.41$ ft in hover conditions and $h_{cg} = -0.32$ ft in cruise conditions. The difference of 0.09 ft (2.7 cm) is due to the fact that derivatives also capture other effects that are not modeled. Constraining the offset in the forward-flight condition to the value identified for hover-flight condition is not indicated, because the offset derivative provides for an additional degree of freedom that is required to properly fit the estimated frequency responses.

Yaw Dynamics Derivatives

Little can be said with regard to the yaw dynamics, since the model structure poorly reflects the physical reality. We had to introduce a constraint between the yaw rate gyro pole $K_{r_{fb}}$ and the natural yaw damping derivative N_r. We can note, however, that the yaw damping N_r has the correct negative sign and that the yaw rate feedback gain K_r stays virtually constant between hover and cruise conditions (2.16 and 2.18 for, respectively, hover and cruise conditions). The major changes are those affecting the yaw control derivative N_{ped}, which decreases, and the lateral speed derivative N_v, which increases drastically in forward flight. Both results are consistent with the expectations. A time delay was identified to account for the high-frequency unmodeled dynamics ($\tau_{ped} =$ 0.099 sec and $\tau_{ped} = 0.059$ sec for hover and cruise conditions, respectively).

Heave Dynamics Derivatives

The heave damping derivative Z_w has the correct sign. In cruise flight, the larger heave damping (65%) and heave control sensitivity Z_{col} (31%) is correct. The latter is directly related to the reduced effect of induced velocity in cruise flight (higher useful power).

3.3 Time Domain Verification

The ultimate model accuracy verification consists of a time domain comparison between the responses recorded during special flight experiments and the model-predicted responses to the recorded pilot inputs time histories. The time domain verification is important for detecting possible modeling shortcomings. A number of uncertainties are introduced throughout the identification process. For example, the estimated frequency responses are already an approximation of the true vehicle dynamics (estimation errors, non-linearities). During the frequency response fit, the model is fitted over a limited frequency range, and the fit is not always perfect. In the time domain verification, these approximations and uncertainties are made visible.

Flight Experiments for Model Validation

The flight data for this verification is collected from special flight experiments. The main idea is to use inputs that are representative of natural inputs used to operate the vehicle. Steps and doublet-like control inputs are typically used in these experiments, because they are commands used during aggressive maneuvering, they can easily be implemented by a pilot, and they have a relatively broad frequency content.

For the Yamaha R-50, we performed the experiments using doublet-like inputs. Experiments involving each pilot input were conducted separately for both the hover and the cruise-flight condition. Each doublet input tries to exploit a large portion of the input deflection available to the pilot. Figures 4.6 and 4.7 show the time response comparison for the hover and cruise condition, respectively. The individual experiments are arranged in columns, for the lateral, longitudinal, collective, and pedal doublet-input experiments, respectively. The first row shows the piloted doublet applied to the respective control input, and the remaining rows show the responses for the vehicle's states. Note that to permit an objective comparison, the axes in the plots were scaled according to the peak response achieved in respective vehicle state. For example, the yaw rate axes covers ± 100 deg/sec, which is the range attained by the yaw rate during the yaw doublet experiment.

Validation Results

Overall, excellent agreement is achieved both in hover and forward flight. We can only find two discrepancies between the responses. Both are in the secondary responses. The first one is in the yaw rate response to the lateral cyclic input in hover flight and is related to the weathervaning effect captured by the N_v derivative. Remember that this derivative was extracted with a high CR bound (32.5%). Better results would require more flight testing. It is possible that the pilot applied too much pedal control during the lateral cyclic sweep to maintain the heading of the helicopter. By closing the loop for the yaw response, this effect was partially masked. The second mismatch occurs for the lateral velocity and acceleration response to the pedal input in cruise flight. This lateral-directional effect is captured by the Y_{ped} derivative, which was also extracted with a relatively high CR bound (21.9%). Here again, to improve the model accuracy, better flight data should be collected.

4. Theoretical Validation of the Identified Derivatives

In the following, we will provide a simple physical interpretation of the key model derivatives, and, at the same time, we will compare their identified values with the values predicted by the underlying physical parameters. This will establish whether the identified model parameters are physically meaningful.

If the model is physically meaningful we can study the effects that the individual modeled components, such as the rotor or the stabilizer bar, have on the overall vehicle dynamics. We can also determine how changes in key parameter values would affect the vehicle's performance and handling qualities. In Chapter 5, we will use the identified model to describe the key characteristics of small-scale rotorcraft, including the effect of scale and the importance of the stabilizer bar, and determine the physical parameters that are key in defining the performance and maneuverability of rotorcraft. Finally, the parameterized model can then also be used to produce an approximation for the dynamics of a vehicle for which we only have basic physical parameters (such as those that can be obtained from directly measurable quantities).

During the development of the parameterized model, we tried to capture the factors that are most relevant for the dynamics of our small-scale helicopter. Because the model is highly simplified, however, its parameters lump together unmodeled effects. By validating the identified parameters through comparison with the values predicted by helicopter theory, we can determine the extent of such approximation.

The key components of our identified model are the main rotor, which produces the main control forces and moments used to control the helicopter, and the stabilizer bar, which acts as a stability augmentation system. The parameters that we want to verify are: the rotor and stabilizer time constants;

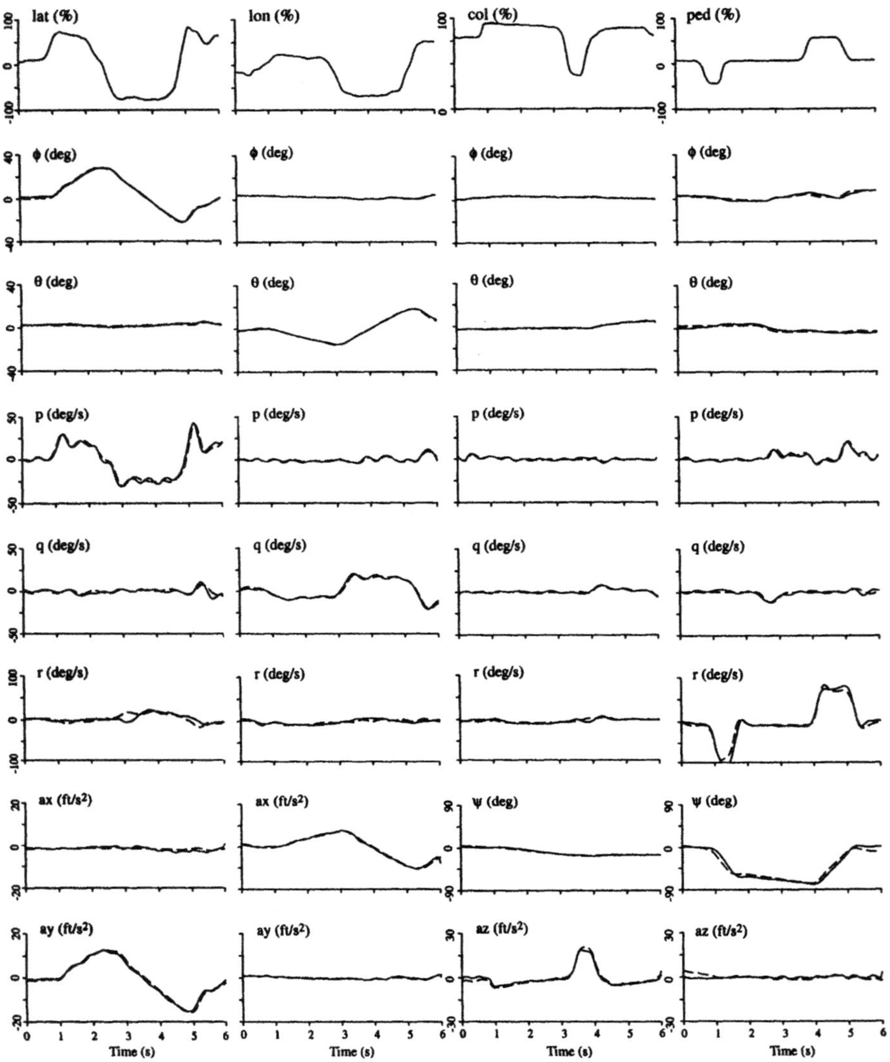

Figure 4.6. Hover flight: comparison between the responses predicted by the identified model (dashed) and the responses obtained during flight testing (solid).

the gearing of the Bell mixer, which couples the stabilizer to the rotor; and the rotor spring derivatives, which capture the coupling of the rotor system to the helicopter body and describe the moments that the rotor can produce. The theoretical determination of the derivatives is based mostly on data that can be determined from measurements taken directly from the vehicle, or using simple

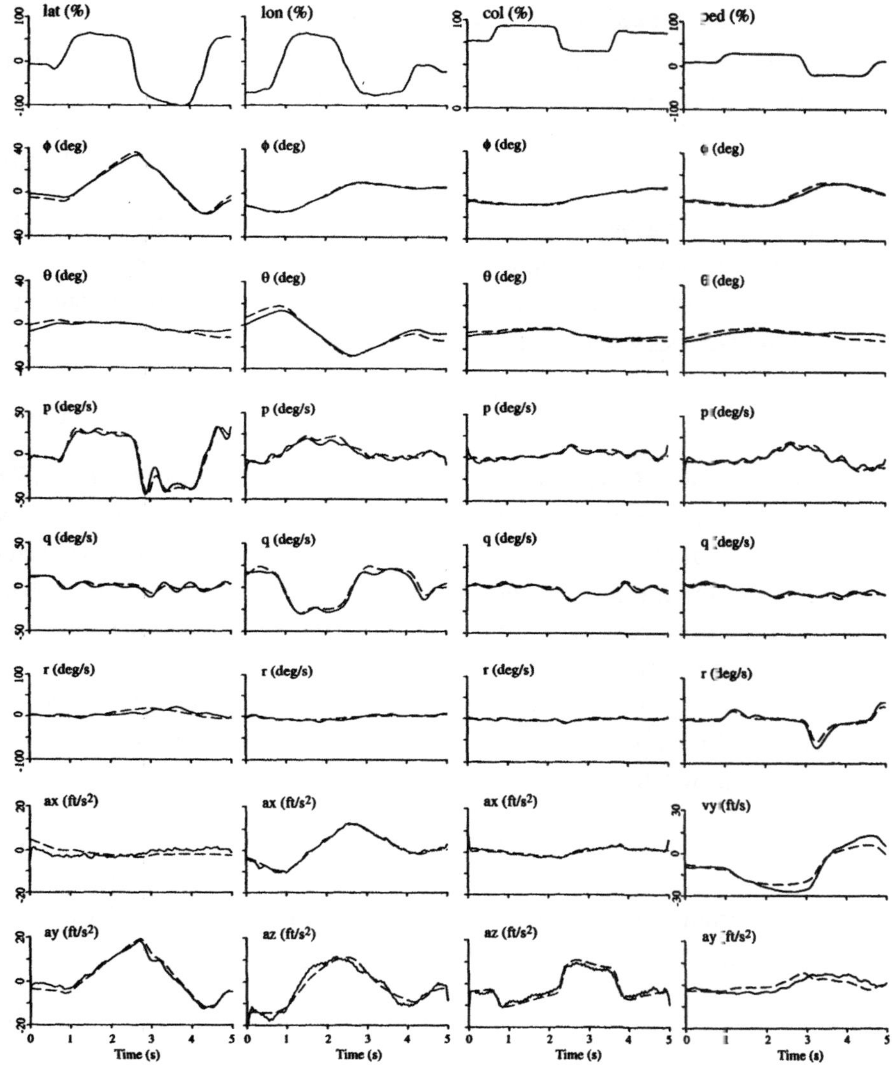

Figure 4.7. Cruise flight ($u_0 = 49$, $v_0 = -11$, and $w_0 = 0$ ft/sec): comparison between the responses predicted by the identified model (dashed) and the responses obtained during flight testing (solid).

experiments. The translational dynamics are important too; however, because of the fuselage aerodynamics, they are not as accessible theoretically (the force damping derivatives and the moment damping derivatives would require wind tunnel testing).

4.1 Rotor and Stabilizer Bar Time Constants

The theoretical value for the rotor time constant was derived earlier from the tip-path-plane equation Eq. 3.58

$$\tau = \frac{16}{\gamma\Omega} \tag{4.19}$$

where γ is the blade's geometrical Lock number defined in Eq. 3.55. The blade moment of inertia I_β was determined using a pendulum experiment.

Since the stabilizer bar's paddles are not continuous from the hub to the tip, we use a modified Lock number formula [92]

$$\gamma = \frac{\rho c C_{l\alpha}(R^4 - r^4)}{I_\beta} \tag{4.20}$$

where R is the radius of the stabilizer bar, r is the inside radius of the paddle (length of the rod), and I_β is the moment of inertia of the stabilizer bar that was calculated from physical data. Moreover, a different lift curve slope needs to be used than for the main blades because of the small aspect ratio (ratio of blade length to chord length) of the paddle. We used the value that was determined experimentally in [92] for a similar type of paddle.

The expression for the rotor time constant is the one for a teetering rotor. The expression for a rotor with a flapping hinge with offset e (measured from the hub) is given by [36]:

$$\tau_f = \frac{16}{\gamma\Omega}\left(1 - \frac{8e}{3R}\right)^{-1}. \tag{4.21}$$

The Lock number can also be modified to include the effect of rotor inflow [36]. The effective Lock number γ^* is defined as:

$$\gamma_{\text{eff}} = \frac{\gamma}{1 + C_{l\alpha}\sigma/16\lambda_i} \tag{4.22}$$

where σ is the rotor solidity (ratio of blades area to rotor disc area), and λ_i is the inflow ratio resulting from a steady uniform inflow velocity v_i. The inflow ratio can be expressed in terms of the rotor thrust coefficient C_T

$$\lambda_i = \frac{v_i}{\Omega R} = \sqrt{\frac{C_T}{2}} \tag{4.23}$$

$$C_T = \frac{T}{\rho\pi R^2(\Omega R)^2}. \tag{4.24}$$

The R-50's tri-hinge system (see Fig. 3.4) is a combination of a teetering hinge and a pair of flapping hinges. In such cases, an equivalent hinge offset

$\tilde{e}$ can be determined and used in place of the geometrical hinge offset. This approach, however, requires a precise knowledge of the rotor restraints. Instead, to give an indication of the range of possible values, we calculated the theoretical rotor time constants for both teetering and hinged configuration.

The values for the parameters used in the calculation of the rotor and stabilizer bar time constants are provided in Table 4.5. With the exception of the main rotor blades' moment of inertia, which was determined through a pendulum experiment, all the parameters were determined from measurements on the vehicle. The results of the comparison between the identified and predicted time constants are shown in Table 4.6. For the hover condition, the identified main rotor time constant agrees strikingly with the theoretical value calculated for a teetering rotor. The values obtained for a hinged rotor are slightly larger. For the stabilizer bar, the agreement is not as good as for the main rotor, but it is sufficiently good to verify its physical meaning.

These results also validate those obtained in our earlier work presented in [60], where the main rotor and the stabilizer bar were modeled using the standard tip-path-plane equation without additional stabilizer bar dynamics. The "effective" rotor time constant that was identified at the time came out to be $\tau_{\text{eff}} = 0.38$ sec. This tells us that the stabilizer bar results in a rotor with an equivalent time constant that is equal to that of the stabilizer bar, i.e. the helicopter behaves as if it had a slower main rotor.

4.2 Parameters of the Bell Mixer

The Bell mixer is a mechanical mixer between the stabilizer bar and the main blade pitch control. The action of this mixer is to impose a command on the main blade pitch that is proportional to the magnitude of the stabilizer bar flapping (see Figure 3.12).

In the rotor-stabilizer equations of motion, the stabilizer bar couples to the main rotor through the coupling derivatives B_d and A_c (Eq. 3.92). These derivatives represent the action of the mechanical mixer. By applying Eq. 3.92, we can calculate the equivalent gain of the mechanical Bell mixer from the identified input derivatives and coupling derivatives:

$$K_d = \frac{B_d}{B_{lat}} = 10.9\ \%/\text{rad} \quad \text{and} \quad K_c = \frac{A_c}{A_{lon}} = 12.9\%/\text{rad}. \tag{4.25}$$

The identification gave us two values, because one parameter was used for each of the lateral and longitudinal flapping equations to provide additional degrees of freedom for the identification. In reality, these values should be the same,

Table 4.5. **Values for the rotor and stabilizer bar physical parameters used for the theoretical calculation of the time constants.**

Stabilizer bar Parameters	
paddle with rod inertia	I_β=0.0349 slug-ft^2
paddle chord length	c=0.33 ft
paddle outside radius	R=1.86 ft
paddle inside radius	r=1.37 ft
paddle lift curve slope	$C_{l\alpha}$=2.67 1/rad
Rotor Parameters	
blade inertia (teetering hinge)	I_β=0.819 slug-ft^2
blade inertia (flapping hinge)	I_β=0.777 slug-ft^2
blade chord length	c=0.351 ft
geometrical hinge offset	e=0.335 ft
rotor radius	R=5.04 ft
blade lift curve slope	$C_{l\alpha}$=5.73 1/rad
General Parameters	
rotor speed	Ω=89.01 rad/sec
rotor thrust	T=150 lbf
rotor solidity	σ=0.0435
air density	ρ=0.00238 slug/ft^3
thrust coefficient	C_T=0.0039
steady inflow component	ν_0=0.0444

Table 4.6. **Comparison between identified and predicted rotor and stabilizer time constants.**

	Identified	Theoretical (teeter)	Theoretical (hinged)
Main rotor (sec)			
Hover	0.046	0.048	0.054
Cruise	0.035	0.035	0.04
Stabilizer bar (sec)			
Hover	0.35	0.38	–
Cruise	0.26	0.33	–

because they correspond to the Bell mixer gearing, which is independent of the position of the blades around the hub.

The theoretical gain of the Bell mixer can be determined from the geometry of the linkage system making up the mixer mechanism. Alternatively, one can also measure the changes in main blade pitch for different amounts of stabilizer bar flapping. From the linkage system geometry we arrived at a value of $K = 13.6$ %/rad which is close to the identified values from Eqs. 4.25. The values of the mixer gains obtained from the derivatives identified in cruise flight are close to the values obtained in hover flight (cruise: $K_d = 10.7$ and $K_c = 13.8$ %/rad). This is an important result, since the mixer gain is a mechanical constant that should also be independent of the flight condition.

The relatively good agreement of the key identified rotor and stabilizer parameters with the values predicted by the physical laws demonstrates that our coupled rotor-stabilizer model produces physically meaningful results and thus is an adequate representation.

4.3 Rotor Moment and Force Derivatives

Recall that the rotor is coupled to the fuselage roll and pitch equations of motion (Eqs. 3.82 and 3.83) through the lateral and longitudinal rotor moment derivatives (flapping spring derivatives) L_b and M_a, respectively, and to the fuselage longitudinal and lateral translational equations of motions (Eqs. 3.79 and 3.80) through the rotor force derivative X_a and Y_b, respectively.

We can calculate theoretical values for these derivatives from the physical definition we introduced in Chapter 3. Recall that the rotor flapping motion produces a moment on the helicopter fuselage due to the hub moments (flapping restraints) and tilting of the thrust vector. The total roll moment acting on the fuselage was given by Eq. 3.31

$$L = L_k + L_T \approx (k_\beta + hT)\, b \qquad (4.26)$$

where k_β is the equivalent torsional spring constant of the center-spring equivalent rotor model, T is the rotor thrust, and hb is the moment arm of the thrust vector with respect to the fuselage center of gravity.

The approximations for the roll and pitch rotor moment (flapping spring) derivatives L_b and M_a, as they are used in the fuselage roll and pitch equations of motion (Eqs. 3.82 and 3.83), are

$$\bar{L}_b \approx \frac{1}{I_{xx}}(Th + k_\beta) \qquad \text{and} \qquad \bar{M}_a \approx \frac{1}{I_{yy}}(Th + k_\beta). \qquad (4.27)$$

The unknown physical parameters are the stiffness of the flapping spring constant k_β, the distance between the rotor hub and the vehicle center of gravity h, and the roll and pitch moments of inertia I_{yy} and I_{xx}, respectively.

The roll and pitch fuselage moments of inertia, as well as the distance between the center of gravity and the rotor head, were determined experimentally from pendulum tests. The values came out as $I_{xx} = 4.41$ slug-ft^2, $I_{yy} = 7.37$ slug-ft^2, and $h = 1.88$ ft. In hover and slow speed, the rotor thrust T is approximately equal to the helicopter weight 150 lbf.

The flapping spring constant k_β was estimated from deflection measurements performed on the rotor. For the Yamaha R-50, the experimental value for the flapping spring constant came out as $k_\beta = 340$ lb-ft/rad. Evaluating the moment derivative equation Eq. 4.27 with the measured values gives us the following values for the roll and pitch flapping spring derivatives: $\bar{L}_b = 140.8$ 1/sec^2 and $\bar{M}_a = 84.2$ 1/sec^2. These results are close to the identified values: $L_b = 166.1$ 1/sec^2 and $M_a = 82.6$ 1/sec^2.

The relative magnitude between the two moment contributions, the stiffness ($k_\beta = 340$ lb-ft/rad) and the thrust ($hT = 280.7$ lb-ft/rad), shows that the rotor stiffness is non-negligible, (i.e. the R-50 rotor is not free to teeter rotor). The flapping spring derivatives are important flight dynamic parameters because they determine the natural frequency of the coupled rotor/fuselage mode and the capability of the rotor to produce control moments.

5. Summary and Final Considerations

The overall results of the identification for the Yamaha R-50 are very good. The state-space model provides for a good fit of the estimated frequency responses, the identified model predicts the time responses of the vehicle to control inputs with high accuracy, and, finally, the key identified derivatives have physical meaning and their value can be related to physical parameters that can be determined from measurements or simple experiments.

The results obtained for our small-scale helicopter are better in terms of frequency response fit and time domain accuracy than what is ordinarily achieved for full-scale helicopters. This brings up the question of whether the dynamics of small-scale rotorcraft are in some ways simpler than those of full-scale rotorcraft. In the following, we will first point out some of the key features of the model structure, which, we recall, is a product of both the model development and the identification process. Then we discuss the validity of the linear model.

5.1 Model Structure

It is interesting to note that these good results were obtained with a model structure that is simpler than the structures that are often necessary to obtain good accuracy for a full-scale model.

In full-scale rotorcraft identification, additional rotor degrees of freedom for coning and lead-lag are often necessary. Moreover, inflow dynamics and engine drivetrain dynamics sometimes need to be accounted for as well. For example, in [89], a coupled coning-inflow model is used for the vertical dynamics; and in [23], an even more comprehensive model, which includes the engine drivetrain dynamics, is used.

The assumptions made for the development of a simplified tip-path-plane rotor model, such as those from [11] presented in Chapter 3, are in many ways more favorable to the characteristics of small-scale rotorcraft than they are to the full-scale ones. For example, the blades of small-scale rotorcraft are highly rigid, and the more agile rotorcraft like the X-Cell also have proportionally faster rotor speed, which reduces the inflow angle. Some of the scaling effects will be explored in the next chapter.

The resulting model for our small-scale rotorcraft has a well-organized block-like structure. There are only cyclic control derivatives for the rotor and stabilizer dynamics, i.e. the cyclic inputs have primary effects only on the rotor and stabilizer bar dynamics. In full-scale models, control derivatives often appear in the angular and translational equations of motion. In our model, the rotor forces and moments propagate through the rigid body dynamics in a physically well defined fashion, i.e. via the rotor forces and moments derivatives (X_a, Y_b, M_a, and L_b).

More experience with small-scale rotorcraft will be necessary to say whether these are qualities that are specific to small-scale helicopters. If this is the case, the model presented herein could be considered a generic small-scale helicopter model. Such a model could greatly facilitate the modeling and analysis of small-scale rotorcraft dynamics. In the following chapter, we will introduce some preliminary identification results obtained with the smaller, more agile X-Cell helicopter. Precisely the same model structure was used; as will be seen in Chapter 5, the results of the attitude model fit are equally good for that helicopter (see Figs. 5.3 and 5.4).

5.2 Validity of the Linear Model

The time domain verification has shown that the linear model accurately predicts the R-50 responses to piloted inputs. However, since we are using a linear model to represent dynamics which are in reality nonlinear, it is important to understand the limitations of such linear models. Nonlinearities are inherent to the physical laws governing the vehicle dynamics. When operating away from the nominal operating point (e.g. hover or particular cruise condition), or when operating at the limits of the operational flight envelope (e.g. saturation of the control inputs, large attitude angles, and high attitude rates), the fidelity of the model is expected to degrade.

With regard to the influence of operating conditions, the results obtained from the time domain verification show that the two models accurately represent the helicopter dynamics for a fair portion of the flight envelope. The hover model was shown to be valid for hover and low speed (up to about 6 m/sec;). The cruise model was identified using flight data covering a broad speed range (10 to 20 m/s), and the time domain validation was performed using flight data that covered about the same speed range. Therefore, a few linear models scattered across the flight envelope would cover large portions of the flight envelope with a level of accuracy that would be sufficient for high-performance control design.

With regard to the domain of aggressive and large amplitude flight, time domain verification shows that the accuracy of the identified linear model is good up to fairly large attitudes and angular rates. For example, the helicopter reaches attitudes of up to 40 degrees for roll and up to 20 degrees for pitch, and the helicopter reaches angular rates of up to 45 deg/sec for roll and up to 30 deg/sec for pitch. To reach these values, the pilot used the entire input range available to him ($\pm 100\%$). Some small-scale helicopters used by hobby pilots, such as the X-Cell, are capable of performing rolls and loops. For such highly maneuverable and aggressive helicopters, additional modeling will need to be performed to cover the extended operating conditions. For example, many maneuvers take place between operating points and thus are by definition outside the linear range. Linear system identification will still be able to produce valuable information about the aircraft dynamics, in particular frequency response identification is still one of the most effective ways to extract accurate information about the dynamics of the vehicle. This will be illustrated in Chapter 5. The extended nonlinear modeling will be summarized in Chapter 7 as part of the ongoing work.

Chapter 5

CHARACTERISTICS OF SMALL-SCALE ROTORCRAFT

Small-scale helicopters have the same basic configuration as their full-scale counterparts. However, because scale operates on the various physical dimensions in different ways, scaling down a helicopter will change the relative magnitude of the main forces governing the vehicle dynamics and thus modify its dynamic characteristics. For the pilot, scaling down a helicopter tends to make it more difficult to control; the stabilizer bar is used to compensate for these scaling effects.

In the following, we will apply principles from dimensional analysis to understand the effects of scale on the flight-dynamic characteristics of helicopters. First, we will outline the key characteristics of the attitude dynamics, applying basic rotorcraft flying qualities metrics and deriving the equivalent metrics for helicopters equipped with a stabilizer bar. We will then show how scaling affects the helicopter performance and flying qualities by applying scaling rules to these metrics. At the same time, we will use dynamic scaling rules to compare the configuration and flight-dynamic characteristics of the Yamaha R-50 with those of a full-scale Bell UH-1H helicopter and those of the more agile X-Cell helicopter. Finally, we will describe the dynamic modes of the R-50 helicopter.

As part of this chapter, we will also introduce some identification results obtained for the X-Cell, which will be used to illustrate the effects of scale and other design features.

1. Characteristics of the Attitude Dynamics

Since any translational motion of a rotorcraft is obtained through rolling or pitching the aircraft, the attitude dynamics are fundamental to the aircraft's entire behavior and performance. The attitude dynamics also determine most of the aircraft's handling qualities characteristics, i.e. how easy it is for the pilot to accomplish a particular task. Piloted vehicles are subject to handling qualities requirements (see [2]). The identified state-space model can be used to evaluate key flying qualities (FQ) metrics. An example of its use to this purpose is presented in [30]. For a comprehensive treatment of this topic see [68]. In the following we will concentrate on the flying qualities metrics that can be derived directly from the state-space equations. Additional considerations on this topic will be made in the next chapter. We will also provide additional insight into the attitude frequency response characteristics and highlight the effects of some key physical parameters, as well as complement the results using identified data from the X-Cell helicopter.

1.1 Attitude Flying Qualities Metrics

Some of the basic FQ metrics can be extracted directly from a quasi-steady representation of the rotorcraft attitude dynamics. The quasi-steady dynamics are obtained by regarding the rotor flapping dynamics as steady, i.e. assuming the rotor reaches a steady flapping much faster than the attitude dynamics settling time. In the following, we first look at the case without stabilizer bar and subsequently develop the expressions for the case with stabilizer bar. All derivations and examples concentrate on the roll dynamics; the results for the pitch dynamics follow from the same reasoning. Also, to keep the expressions concise we discard the cross-coupling terms. Cross-axis parameters can be derived similarly (see [30]).

Without Stabilizer Bar

For a plain rotor system, the quasi-steady attitude dynamics are obtained by setting the flapping time derivative $\dot{b}$ to zero in the flapping equations of motion Eq. 3.57

$$\dot{b} = 0 = -b - \tau_f p + B_{lat}\delta_{lat}. \tag{5.1}$$

The steady-state flapping is

$$b = -\tau_f p + B_{lat}\delta_{lat}. \tag{5.2}$$

The quasi-steady roll dynamics are obtained by substituting the steady flapping in the roll equation of motion Eq. 3.82

$$\dot{p} = -L_b\tau_f p + L_b B_{lat}\delta_{lat}. \tag{5.3}$$

The key flying qualities parameters are

- The effective *roll damping* derivative ($\frac{1}{sec}$) is defined as:

$$(L_b)_e = -L_b\tau_f \tag{5.4}$$

- The effective *roll control* derivative ($\frac{rad/sec^2}{\%}$) (also called *control sensitivity* because it determines the initial roll acceleration [68]) is defined by:

$$(L_{\delta_{lat}})_e = L_b B_{lat}. \tag{5.5}$$

- The *roll rate sensitivity* ($\frac{rad/sec^2}{\%}$) is defined as the ratio of control sensitivity to roll damping:

$$\frac{p}{\delta_{lat}}\bigg|_{s=0} = \frac{L_{\delta_{lat}}}{(L_b)_e} = \frac{L_b B_{lat}}{L_b\tau_f} = \frac{B_{lat}}{\tau_f} \tag{5.6}$$

The roll rate sensitivity gives the steady-state roll rate that can be attained by the helicopter (per percent of lateral cyclic stick). It can also be derived from the roll rate transfer function:

$$\frac{p}{\delta_{lat}} = \frac{L_b B_{lat}}{s + L_b\tau_f}, \tag{5.7}$$

from which we obtain the *roll rate time constant* (sec):

$$\tau_p = \frac{1}{L_b\tau_f}. \tag{5.8}$$

With Stabilizer Bar

Typically, the stabilizer bar time constant τ_s is several times larger than the rotor time constant τ_f. Hence, when deriving the quasi-steady attitude dynamics for the case with a stabilizer bar, we can proceed in two stages. First, we consider the main rotor dynamics as steady $\dot{b} = 0$ and still account for the stabilizer bar dynamic. Second, we also regard the stabilizer bar as steady.

The augmented lateral flapping equation of motion was introduced in Eq. 3.91; using the form with the stabilizer bar coupling derivative B_d we have

$$\tau_f\dot{b} = -b - \tau_f p + B_d d + B_{lat}\delta_{lat}. \tag{5.9}$$

In steady state (setting $\dot{b} = 0$) we have

$$b = -\tau_f p + B_d d + B_{lat}\delta_{lat}. \tag{5.10}$$

To couple the stabilizer dynamics we derive the transfer function for the stabilizer bar flapping dynamics by applying the Laplace Transform to Eq. 3.87

$$\tau_s s d(s) = -d(s) - \tau_s p(s) + D_{lat}\delta_{lat}(s). \tag{5.11}$$

We get the following transfer function for the stabilizer bar response

$$d(s) = \frac{-\tau_s p(s) + D_{lat}\delta_{lat}(s)}{\tau_s s + 1}. \tag{5.12}$$

Inserting $d(s)$ in the steady rotor flapping equation 5.10 we obtain

$$b(s) = -\tau_f p(s) + B_d \left(\frac{-\tau_s p(s) + D_{lat}\delta_{lat}(s)}{\tau_s s + 1} \right) + B_{lat}\delta_{lat}(s) \tag{5.13}$$

To derive the first-order differential equation for the main rotor flapping under the influence of the stabilizer bar dynamics, we proceed by an inverse Laplace transform of Eq. 5.13. To be consistent with our assumption that the rotor dynamics are steady, we set the time differential $\dot{p}$ and $\dot{\delta}_{lat}$ equal to zero and obtain:

$$\tau_s \dot{b} = -b - (\tau_f + B_d\tau_s)p + (B_{lat} + B_d D_{lat})\delta_{lat}. \tag{5.14}$$

We see that this equation is identical to the original rotor flapping equation 3.91, with the exception that: the main rotor time constant has been replaced by the stabilizer bar time constant τ_s; the rotor lags by an additional $B_d\tau_s$ due to the stabilizer bar contribution; and the cyclic control is augmented by an amount $B_d D_{lat}$. We refer to this model as the lumped rotor-stabilizer model, which can be regarded as an effective rotor dynamics model, where the stabilizer bar dominates the dynamic response of the rotor.

The lumped formulation also explains that when using the standard coupled rotor-fuselage (R-F) model structure for the identification of a helicopter with a stabilizer bar, we identify an effective rotor time constant which is close to the stabilizer bar time constant. The remaining derivatives lump the damping and command boosting effects of the stabilizer bar (the identification results using this lumped model were presented in [60]).

The next step involves regarding the entire rotor system in steady state. Setting $\dot{b}$ to zero in Eq. 5.14 we obtain the following steady-state flapping response

$$b = -(\tau_f + B_d\tau_s)p + (B_{lat} + B_d D_{lat})\delta_{lat}. \tag{5.15}$$

Using this expression in the roll dynamics equation we obtain the following quasi-steady roll dynamics for the case with a stabilizer bar

$$\dot{p} = -L_b(\tau_f + B_d\tau_s)p + L_b(B_{lat} + B_d D_{lat})\delta_{lat}. \tag{5.16}$$

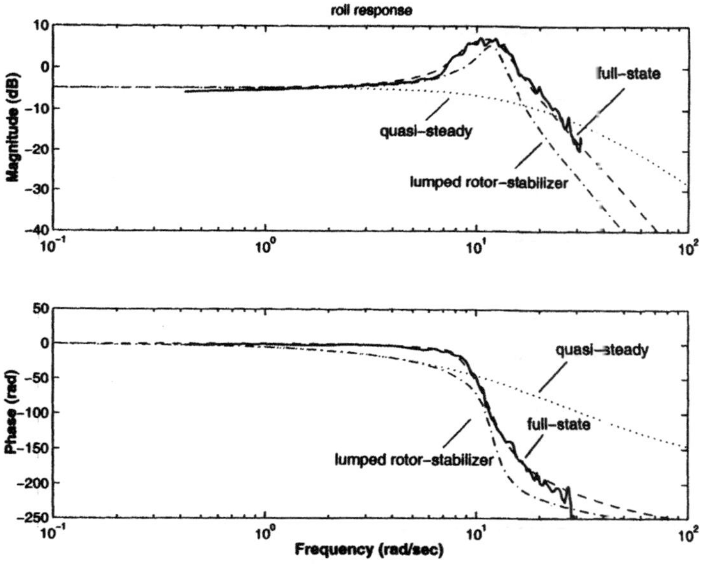

Figure 5.1. **Comparison between the estimated R-50 roll frequency responses and the responses from the identified full-order attitude model. Also shown are the lumped model and quasi-steady approximations.**

Comparing the equation of motion to Eq. 5.3 we can see the additional terms due to the stabilizer bar. We will explain their contributions next.

Figures 5.2 and 5.1 show the comparison between the full-order model, the lumped model and the quasi-steady attitude models, for the roll and pitch axes, respectively. The contribution of the stabilizer bar to the frequency response, i.e. the relevance of the augmented model, is most noticeable in the pitch response magnitude and phase in the 10 to 20 rad/sec frequency range. The lumped model does not capture this detail as precisely as the full-order attitude model. The quasi-steady model, incapable of capturing the second-order characteristic of the rate response, is only valid up to about 2-3 rad/sec. Note that the two reduced-order attitude dynamics models can also be used with the translational dynamics.

For the FQ metrics for the rotor with stabilizer bar we proceed from Eq. 5.16 and determine the same flying qualities parameters as for the plain rotor

- The effective *roll damping* derivative ($\frac{1}{sec}$) is defined as:

$$(L_b)_e = -L_b(\tau_f + B_d\tau_s) \tag{5.17}$$

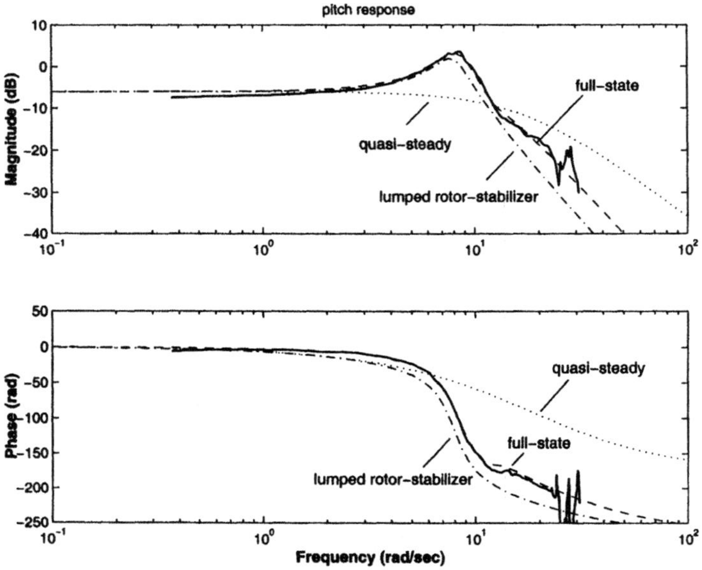

Figure 5.2. Comparison between the estimated R-50 pitch frequency responses and the responses from the identified full-order attitude model. Also shown are the lumped model and quasi-steady approximations.

- The effective *roll control* derivative $(\frac{rad/sec^2}{\%})$ is defined by:

$$(L_{\delta_{lat}})_e = L_b(B_{lat} + B_d D_{lat}). \tag{5.18}$$

- The *roll rate sensitivity* $(\frac{rad/sec^2}{\%})$ is defined as the ratio of control sensitivity to roll damping:

$$\frac{p}{\delta_{lat}}\Big|_{s=0} = \frac{L_{\delta_{lat}}}{(L_b)_e} = \frac{L_b(B_{lat} + B_d D_{lat})}{L_b(\tau_f + B_d \tau_s)} = \frac{B_{lat} + B_d D_{lat}}{\tau_f + B_d \tau_s} \tag{5.19}$$

The quasi-steady roll rate transfer function is

$$\frac{p}{\delta_{lat}} = \frac{L_b(B_{lat} + B_d D_{lat})}{s + L_b(\tau_f + B_d \tau_s)}, \tag{5.20}$$

from which we obtain the *roll rate time constant* (sec):

$$\tau_p = \frac{1}{L_b(\tau_f + B_d \tau_s)}. \tag{5.21}$$

Notice the additional terms involving the stabilizer bar gearing parameter B_d, the stabilizer bar time constant τ_s, and the stabilizer bar cyclic input derivative D_{lat}. In particular: the increase in roll damping; the boosting effect in the effective roll control derivative, which can be used to relieve the servo actuators (or pilot); and the compound effect on the rate sensitivity. Through adjustments to the stabilizer bar parameters, significant changes to the FQ can be achieved. We will see later in this chapter how this can be used to offset the effects of scaling down the size of a helicopter. The parameters for both the plain rotor and the rotor with stabilizer bar are summarized in Table 5.4.

1.2 Attitude Rate Transfer Function

As seen earlier in Figs. 5.3 and 5.4, the characteristics of the attitude rate dynamics are very close to those of a second-order system. To better understand the effect of the key physical parameters and stabilizer bar on the rate response characteristics, we will use the rate response transfer functions.

For a plain rotor, the transfer function describing the lateral rotor flapping response to the lateral cyclic input and the helicopter roll rate is

$$b = \frac{-p}{s + 1/\tau_f} + \frac{B_{lat}\delta_{lat}}{\tau_f(s + 1/\tau_f)}. \tag{5.22}$$

After inserting this expression into the fuselage roll equation ($\dot{p} = L_b b$) and solving for the roll rate, we get the following transfer function for the helicopter roll response to lateral cyclic input:

$$\frac{p}{\delta_{lat}} = \frac{L_b B_{lat}/\tau_f}{s^2 + (1/\tau_f)s + L_b} \tag{5.23}$$

As expected, the characteristic equation for the roll rate response is a second-order polynomial. Its roots and steady-state gain fully characterize the attitude dynamics. The polynomial has one pair of conjugate complex roots that corresponds to the coupled R-F mode. The roots of the characteristic equation can be expressed in terms of the model parameters. The natural frequency of the coupled R-F mode is given by the flapping spring derivative L_b:

$$\omega_{\text{roll-flap}} = \sqrt{L_b}, \tag{5.24}$$

and the associated damping ratio is:

$$\zeta_{\text{roll-flap}} = 1/(2\tau_s\sqrt{L_b}). \tag{5.25}$$

For the case with a stabilizer bar the rate response has a third-order characteristic. The expressions for its roots are too complicated to be insightful. However, to simplify the problem, we can approach it by considering the lumped rotor model for the case with the stabilizer bar (see Eq. 5.14). Based on Figs. 5.2 and 5.1, we see that the lumped rotor-stabilizer model provides a relatively good approximation of the characteristic of the attitude rate response. The roll rate transfer function based on the lumped rotor model is

$$\frac{p}{\delta_{lat}} = \frac{L_b\left(B_{lat} + B_d D_{lat}\right)/\tau_s}{s^2 + (1/\tau_s)s + L_b\left(\tau_f + B_d\tau_s\right)/\tau_s}. \tag{5.26}$$

Equation 5.25 shows that the small damping ratio of the coupled R-F modes directly reflects the large equivalent rotor time constant. For example, for the R-50, we have $\zeta_{\text{roll-flap}} = 1/(2\tau_s\sqrt{L_b}) = 0.11$ and $\zeta_{\text{pitch-flap}} = 1/(2\tau_s\sqrt{M_a}) = 0.16$ for, respectively, the roll and pitch axes.

1.3 Key Physical Parameters

From the basic flying qualities metrics we can recognize the physical parameters that are determinant for the attitude dynamics. A high rate sensitivity, i.e. high maximum roll (or pitch rate), is enabled by

- small rotor time constant τ_f; or effective time constant with stabilizer bar ($\tau_{\text{eff}} \approx \tau_f + B_d\tau_s$)
- high cyclic control authority; or effective authority with stabilizer bar ($B_{lat,\text{eff}} \approx B_{lat} + B_d D_{lat}$)

A fast transient rate response, i.e. high roll acceleration is enabled by

- high hub stiffness and small vehicle moment of inertia, which are both captured by L_b
- small rotor damping moment, i.e. small rotor time constant; or effective time constant
- large cyclic authority (maximum blade pitch deflection; possibility for blade stall), i.e. large B_{lat}

1.4 Identified X-Cell Attitude Dynamics

To illustrate the characteristics of small-scale rotorcraft, we include identification results obtained from MIT's X-Cell helicopter. Table 5.1 shows the attitude dynamics derivatives identified for the X-Cell. We also show the values obtained for the R-50 and their ratio with respect to the X-Cell.

Figures 5.3 and 5.4 show the estimated and model predicted roll and pitch rate frequency responses obtained for both the R-50 and the X-Cell helicopter.

Table 5.1. **Identified Yamaha R-50 and X-Cell .60 parameters in hover (†these derivatives were set to zero).**

	X-Cell	R-50	ratio
Angular Dynamics Parameters			
τ_f	0.052	0.046	0.88
τ_s	0.22	0.34	1.55
L_b	320	166	0.52
M_a	204	82.6	0.40
A_b	$0^\dagger$	-0.189	–
B_a	$0^\dagger$	0.368	–
A_c	1.15	0.644	0.56
B_d	1.15	0.71	0.62
A_{lat}	$0^\dagger$	0.015	–
A_{lon}	0.53	0.05	0.09
B_{lat}	0.42	0.07	0.17
B_{lon}	$0^\dagger$	-0.007	–
C_{lon}	0.11	0.13	1.18
D_{lat}	0.11	0.135	1.22

Notice the difference between the two helicopters, in the rate sensitivity and the phase bandwidth. Also visible is the difference in the natural frequency of the coupled R-F mode. In the following we will provide some insight into the origin of these differences by studying the effects of scale on the flight-dynamic characteristics.

2. Scaling Laws

In the following we will introduce scaling laws that will help us explain some of the effects scale has on the basic helicopter dynamic characteristics. With such scaling laws we should be able to form a more precise idea of the performance that we can expect from small-scale rotorcraft, and how some key flying qualities parameters are affected by scale. Scaling rules will also help us in performing meaningful comparison between helicopters of different sizes.

Scale affects the dynamic characteristics of a vehicle (or other system) by changing the relative magnitude of the governing forces (e.g. inertial, gravita-

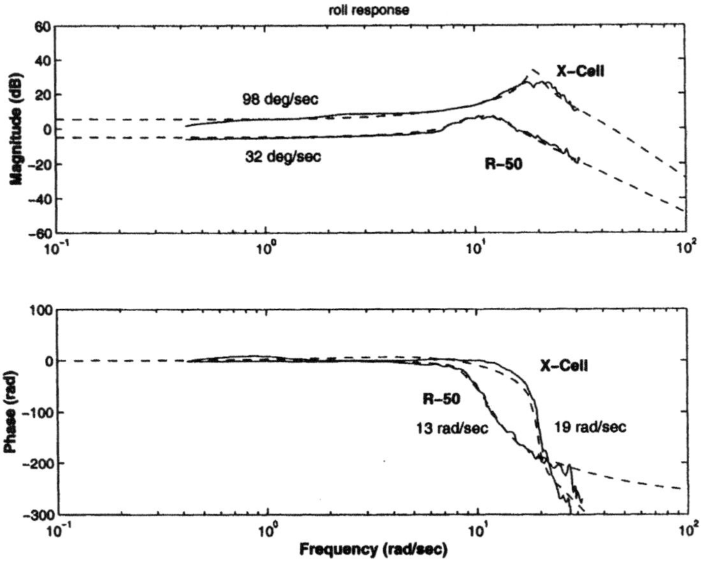

Figure 5.3. Comparison between the estimated R-50 and X-Cell roll frequency responses showing the rate sensitivity and phase bandwidth.

tional, aerodynamic). These scaling effects basically depend on how the underlying physical dimensions and parameters change with scale. Physical laws impose partial constraints on how some physical dimensions change with the length dimension. For example, let a scale factor N refer to a model helicopter with $1/N$ the rotor diameter of some prototype. If we assume that the helicopter stays geometrically similar (all helicopter dimensions change inversely proportional to N), and that the material density is unchanged, the weight will scale by a factor $1/N^3$, while the moments of inertia will scale by a factor $1/N^5$. Some vehicle dimensions, like its mass or moments of inertia, are strongly influenced by size, others, usually those considered as design parameters (e.g. rotor speed or rotor stifness), can more easily take values independently of scale.

In the following we will investigate the application of two different scaling hypotheses: Froude scaling and Mach scaling. Each imposes a particular set of scaling rules on the physical parameters and results in particular dynamic scaling. Subsequently, we will see if the two scaling hypotheses are meaningful descriptions by applying them to a range of differently sized helicopters.

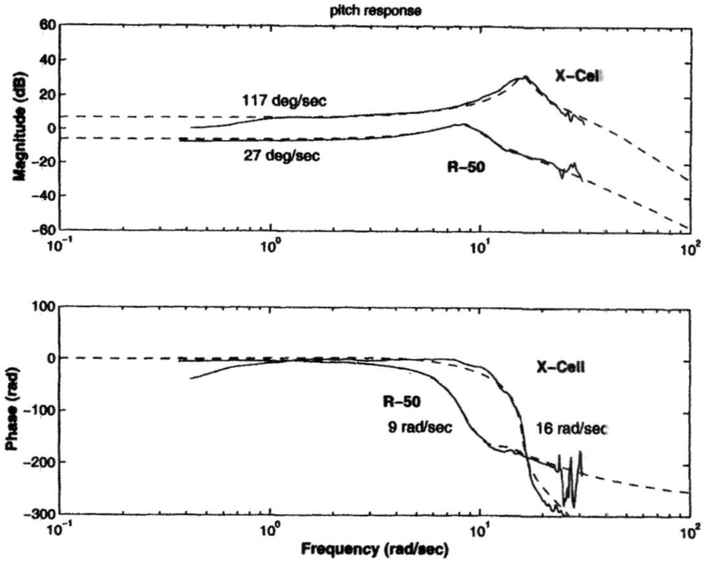

Figure 5.4. **Comparison between the estimated R-50 and X-Cell pitch frequency responses showing the rate sensitivity and phase bandwidth.**

2.1 Froude Scaling

The first scaling hypothesis is that rotorcraft are designed to be dynamically similar to some prototype. A model is said to be dynamically similar if the relative magnitude of the governing forces in the system are unchanged by scale. One form of dynamic similarity is found in Froude models which have the same ratio of inertia to gravitational forces. This ratio is captured by the Froude number

$$\boldsymbol{F} = \frac{V^2}{Lg} \tag{5.27}$$

where V is a characteristic velocity, L is a characteristic length, and g is the gravitational acceleration.

Froude models are often used to work with experiments at reduced scale [51]. An example where a reduced-scale rotorcraft model was used for the study of basic stability and control characteristics is available in [17]. Another such example, this time where a reduced-scale airplane was used to study handling qualities, is provided in [8].

To find the scaling rules that help enforce dynamics similarity between two vehicles, we can start with the Froude number; we choose the rotor tip speed

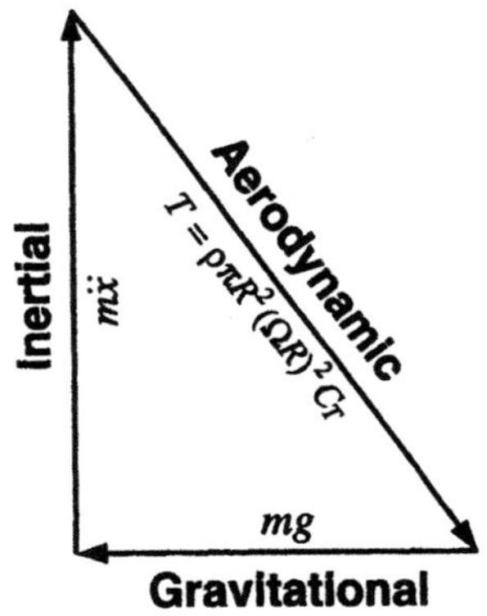

Figure 5.5. Representation of the relation between inertial, gravitational, and aerodynamic forces. Two aircraft will be dynamically similar if their respective force triangles have the same proportion.

ΩR as the characteristic velocity (since it determines the speed envelope of the vehicle) and the rotor radius as the characteristic length. Enforcing the same Froude number for the model (subscript m) and the prototype (subscript p), knowing that $R_m = R_p/N$, requires

$$\Omega_m = \sqrt{N}\Omega_p \tag{5.28}$$

Note that the relation of the Froude number also imposes a relation between time scales, namely

$$\sec \sim \frac{1}{\sqrt{N}} \tag{5.29}$$

Scaling rules for the various physical dimensions follow directly from the geometrical constraints (assuming constant material density) and time scale relation.

For an aircraft, the governing forces are the inertial forces, the gravitational forces, and the aerodynamic forces. Figure 5.5 shows these forces put in relation to each other to form a triangle. The expression for the rotor thrust T is used to represent the aerodynamic forces. Using this representation, we can say that a scaled model will be dynamically similar to its prototype if the proportion of the force triangle is unchanged by scale.

To make a model dynamically similar to a prototype system, its physical dimensions and parameters must be scaled in a way that preserves the ratio among these governing forces. Froude scaling takes care of the ratio of inertial to gravitational forces. At this point, to complete dynamic similarity, there remains the ratio of either aerodynamic to inertial $R_{A/I}$, or aerodynamic to gravitational $R_{A/G}$. Note that it is not necessarily possible to have complete dynamic similarity.

We choose to look at $R_{A/G}$ using the thrust-to-weight ratio expressed in terms of the rotor thrust coefficient C_T

$$R_{T/W} = \frac{T}{W} = \frac{\rho\pi\Omega^2 R^4 C_T}{mg}, \tag{5.30}$$

where m is the vehicle mass. Given that two dynamic similar rotorcraft already have similar Froude numbers and thus rotor speeds related by Eq. 5.28, we deduce that a similar $R_{T/W}$ (i.e., similar $R_{A/G}$) requires

$$C_{T,m} = C_{T,p} \tag{5.31}$$

The thrust coefficient C_T is a function of the blade geometry and airfoil characteristic, but also of the Mach number (airspeed) and the Reynolds number. A Similar thrust-to-weight coefficient together with a similar Froude number completes the requirements for dynamic similarity.

2.2 Mach Scaling

Clearly, all rotorcraft are not dynamically similar, and we need additional ways to relate helicopter behavior across scale. A useful consideration is main rotor tip speed, which places an upper limit on a vehicle's forward velocity. This restriction exists since high translational velocities can lead to excessive air speed on the advancing rotor blade (and therefore flow compressibility) and insufficient speed on the retreating rotor blade (leading to blade stall). As such, if two vehicles at different scales have similar translational flight envelopes, they are expected to have similar rotor tip speeds. Also, given the difference in scale, it is then reasonable to expect a significant dissimilarity in overall dynamic behavior.

Consider two helicopters, with relative scale N but equal tip speed, i.e.

$$\Omega_p R_p = \Omega_m R_m. \tag{5.32}$$

Since their rotor radii follow $R_m = R_p/N$, we must have

$$\Omega_m = N\Omega_p, \tag{5.33}$$

and also the following relation between time constants:

$$T_m = \frac{1}{N} T_p. \tag{5.34}$$

Table 5.2. Froude and Mach Scaling Rules.

Dimension	Froude	Mach
Length	$L_m = L_p/N$	$L_m = L_p/N$
Time constant	$T_m = T_p/\sqrt{N}$	$T_m = T_p/N$
Speed	$V_m = V_p/\sqrt{N}$	$V_m = V_p$
Weight	$W_m = W_p/N^3$	$W_m = W_p/N^3$
Moment of inertia	$I_m = I_p/N^5$	$I_m = I_p/N^5$
Frequency	$\omega_m = \omega_p\sqrt{N}$	$\omega_m = \omega_p N$

Comparing the rotor speed constraint of Eq. 5.28 to Eq. 5.32, we note that vehicles with similar tip speeds (said to exhibit "Mach similarity") cannot be dynamically similar. In fact, the thrust-to-weight ratio of a Mach model will change according to

$$R_{T/W,m} = N R_{T/W,p} \tag{5.35}$$

The next sections will show that Mach scaling represents a significant departure from dynamic similarity and helps explain the extreme agility of some small-scale rotorcraft.

Table 5.2 gives Froude and Mach scaling rules for other physical quantities, again assuming the model and prototype have equal vehicle densities and operate in equal-density air.

2.3 Scaling Hypotheses

In the following, we will test our two scaling hypotheses by comparing the model characteristics, i.e. Froude number, Mach number, and thrust-to-weight ratio, for differently sized rotorcraft. We will determine if some of these rotorcraft share similar model characteristics. If this proves to be the case, we will be able to use the two sets of scaling laws to determine the effect of scale on their dynamics characteristics and relate differently sized rotorcraft.

For the comparison, we choose two full-size helicopters: the Bell UH-1H and the Robinson R22, and our two small-scale helicopters: the Yamaha R-50 and the X-Cell (for both the nominal and instrumented configuration). Their main physical dimensions are presented in Table 5.3 along with the scale ratio N (calculated relative to the R-50), the Froude and Mach numbers, and the

Table 5.3. **Comparison of key physical and model characteristics of various helicopters with scale ratio relative to the R-50.**

	Physical Characteristics			Computed Characteristics			
	Weight lb	Radius ft	Speed rad/sec	Scale N	Froude Number	Mach Number	$\frac{\text{Thrust}}{\text{Weight}}$
UH-1H	8000	24	34	0.21	862	0.75	0.94
R22	1300	13	53	0.38	1134	0.63	1.21
R-50	97	5	89	1	1230	0.41	1
CMU's	150	5	89	1	1230	0.41	0.65
X-Cell	10	2.5	167	2	2165	0.38	2.13
MIT's	17	2.5	167	2	2165	0.38	1.25

thrust-to-weight ratios. Notice that the scales of these helicopters span a full order of magnitude ($0.21 \leq N \leq 2$).

From the results, we see that the UH-1H, the R22, and the R-50 share similar Froude numbers and thrust-to-weight ratios (CMU's R-50 is lower due to the heavy onboard instrumentation). The calculated hover thrust coefficients for these helicopters are shown later in Table 5.8. Here again the first three helicopters share similar values. Note that on the other hand their Mach numbers are significantly different. Based on these observations we can say that the UH-1H, R22, and R-50 share some degree of dynamic similarity.

Looking at the X-Cell, we notice that it has a Froude number which is significantly different from the three other helicopters. The same is true for its thrust-to-weight ratio and its thrust coefficient. Its Mach number, on the other hand, is very close to that of the R-50, indicating that the Mach scaling rules may help relate these two vehicles.

As a further application and illustration of the scaling laws, we plot the thrust-to-weight ratio $R_{T/W}$ obtained for the different helicopters in a diagram showing the effect of Froude and Mach scaling on that ratio as well as a third form of scaling: constant rotor speed. For the comparison, we set the thrust coefficient for all helicopters equal to the value that was calculated for the uninstrumented R-50 in hover flight (C_T=0.00262). And all thrust-to-weight ratios were then normalized with respect to the R-50's value. Figure 5.6 shows that the UH-1H, the R22, and the uninstrumented R-50 all lie along a line of similar $R_{T/W}$, while the uninstrumented X-Cell clearly escapes this group of

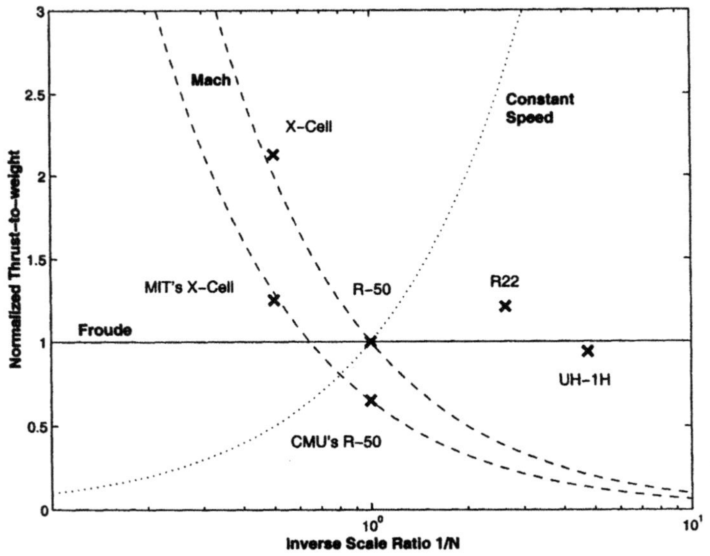

Figure 5.6. Normalized thrust-to-weight ratio for different full-scale and model-scale helicopters. The helicopters are arranged in function of their size relative to the R-50. Also shown is the effect of scale on the thrust-to-weight ratio resulting from different rotor speed scalings rules (solid: Froude; dashed: constant blade-tip speed; dotted: constant rotor speed).

vehicles. In fact, the latter relates nicely to the R-50 through the Mach scaling curve. This is shown for both the instrumented MIT and CMU test vehicles as well as for their nominal configuration.

Figure 5.7 shows the heave acceleration to collective input frequency response estimated from the flight data. At 1 rad/sec, the R-50 exhibits a vertical acceleration of about 11 m/sec^2 (1.1g), while the X-Cell's value is about 21 m/sec^2 (2.1g). The factor two difference agrees with the ratios depicted in Fig. 5.6 and with the Mach scaling rule for the thrust-to-weight ratio shown in Eq. 5.35.

These results suggest that we can divide our six helicopters into two classes. The class of dynamically similar "traditional" helicopters. And a class of more aggressive "aerobatic" helicopters. The results also show that Froude and Mach scaling principles could be a reasonable way of describing the effect of scale on the dynamic characteristics as well as to relate the characteristics of the helicopter within and across the two classes.

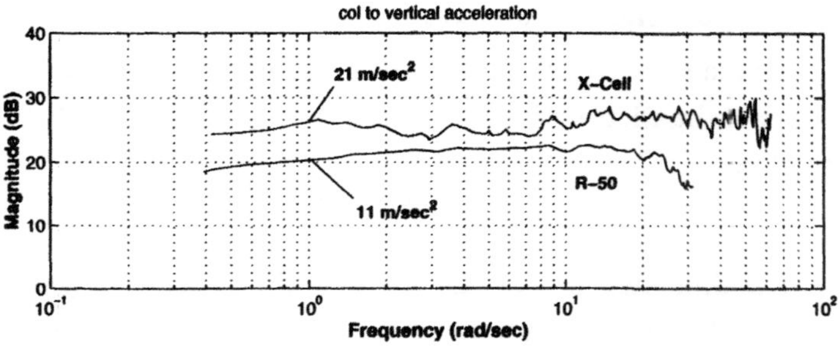

Figure 5.7. Identified frequency responses of heave acceleration to collective input for the R-50 and X-Cell helicopters.

The thrust-to-weight ratio accounts for only part of the helicopter's performance and dynamic characteristics. In the following, we will first apply Froude and Mach scaling to determine the effect of scale on the helicopter's attitude dynamics and flying qualities. Subsequently, we will apply scaling rules to compare the characteristics of the R-50, X-Cell and Bell UH-1H helicopters in more detail.

3. Effects of Scale on Rotorcraft Dynamics

Small-scale helicopters are known to have a higher control sensitivity and faster angular dynamics than full-scale helicopters. In the following, we will use the scaling relations to demonstrate this fact. First, we investigate the effects of scale on the key physical parameters relevant to the attitude dynamics, and then on the flying qualities metrics introduced earlier.

3.1 Effects of Scaling on the Key Physical Parameters

The helicopter's attitude dynamics are central to the overall vehicle dynamics and flying qualities. To determine the effects of scale on these dynamics, we will study the transfer functions derived earlier from the coupled R-F Eqs. 5.3 and 5.16. Note that we could also directly apply the Froude and Mach scaling rules based on the units of the respective dimensions. However, for a better physical insight into the effects of scale, we will determine how scaling affects the key underlying physical dimensions enumerated earlier in section 1.3.

Effect of Scaling on the Rotor Moments

We can determine the effect of scaling on the moment produced by the rotor by using the approximation derived in Chapter 4 (see Eq. 4.27) for the rotor flapping derivative

$$\bar{L}_b \cong \frac{1}{I_{xx}}(Th + k) \tag{5.36}$$

To simplify the scaling analysis, we assume that the relative magnitude of the thrust and rotor stiffness contributions to the moment remain equal with scale.

For a Froude model, the thrust-to-weight ratio $R_{T/W}$ stays constant, thus, the thrust T changes proportionally to the helicopter weight (N^{-3}). For a Mach model, the thrust-to-weight ratio will increase (N^{-2}) as the scale is reduced. The distance between the hub and the center of gravity h scales by N^{-1}, and the fuselage moment of inertia I_{xx} scales by N^{-5}. The compounded effect for the Froude model is:

$$L_b|_m = \frac{1}{N^{-5} I_{xx}}(N^{-3}TN^{-1}h + N^{-4}k) = NL_b|_p, \tag{5.37}$$

which shows that the flapping spring derivative L_b scales by a factor N (N^2 for the Mach model).

By applying these results to the natural frequency of the coupled R-F mode (Eq. 5.24), we get:

$$\omega_{\text{roll-flap}}|_m = \sqrt{N}\sqrt{L_b} = \sqrt{N}\omega_{\text{roll-flap}}|_p. \tag{5.38}$$

for the Froude model ($\omega_{\text{roll-flap}}|_m = N\omega_{\text{roll-flap}}|_p$ for the Mach model). We can derive the same results (5.37 and 5.38) by inspecting the units of the dimensions. For example, the rotor moment has units $1/\text{sec}^2$; based on Froude's time scaling (Eq. 5.29), it scales by N. The frequency has units rad/sec and therefore scales by $\sqrt{N}$. The same analysis can be performed for the Mach model.

Effect of Scaling on the Rotor Time Constant

Remember that the rotor time constant for a teetering rotor is a function of the blade lock number γ and the rotor speed Ω (Eq. 4.19). The lock number actually represents the ratio of aerodynamic to inertial forces. Thus, for a Froude model it should not change with scale. This conclusion can also be reached knowing that γ is a dimensionless parameter.

Based on the expression for the rotor time constant (Eq. 4.19), and the fact that for a Froude model the rotor speed scales by the square root of the scale factor ($\Omega|_m = \sqrt{N}\Omega|_p$), we determine that the rotor time constant scales as follows:

$$\tau|_m = \frac{1}{\sqrt{N}}\frac{16}{\gamma\Omega_p} = \frac{1}{\sqrt{N}}\tau|_p. \tag{5.39}$$

For the Mach model, the time constant would scale by a factor $1/N$. Same as above, these scalings can be derived from the units of the time constant: the time constant is in seconds, and hence by Eq. 5.29 scales as $1/\sqrt{N}$ for the Froude model and as $1/N$ for the Mach model.

With that, we can conclude that scaling does not affect the damping of the coupled R-F mode for either scaling law:

$$\zeta_{\text{roll-flap}}|_m = \frac{1}{2(\tau_s/\sqrt{N})\sqrt{N}\sqrt{L_b}} = \zeta_{\text{roll-flap}}|_p. \tag{5.40}$$

3.2 Effect of Scaling on Basic Flying Qualities

The effect of scaling on the flying qualities parameters is determined by the effects of scale on the rotor time constant and flapping spring derivative (or directly from the units):

- The effective *roll damping* derivative becomes:

$$(L_b)_e|_m = -L_b|_m\tau_f|_m = -NL_b|_p\frac{1}{\sqrt{N}}\tau_f|_p = \sqrt{N}(L_b)_e|_p. \tag{5.41}$$

- The *control sensitivity* becomes:

$$(L_{\delta_{lat}})_e|_m = L_b|_m B_{lat} = NL_b|_p B_{lat} = N(L_{\delta_{lat}})_e|_p. \tag{5.42}$$

- The roll *rate sensitivity* becomes:

$$\frac{p}{\delta_{lat}}\Big|_{s=0}\Big|_m = \frac{B_{lat}}{\tau_f|_m} = \frac{\sqrt{N}B_{lat}}{\tau_f|_m} = \sqrt{N}\frac{p}{\delta_{lat}}\Big|_{s=0}\Big|_p. \tag{5.43}$$

- The roll *rate response time constant* becomes:

$$\tau_p|_m = \frac{1}{L_b|_m\tau_f|_m} = \frac{\sqrt{N}}{N}\frac{1}{L_b|_p\tau_f|_p} = \frac{1}{\sqrt{N}}\tau_p|_p. \tag{5.44}$$

Table 5.4 summarizes the Froude and Mach scaling effects on the attitude flying qualities metrics along with the respective expressions for a helicopter with and without stabilizer bar. We observe that, when scaling down a helicopter (increasing N), its control sensitivity increases proportionally to the scale ratio for the Froude model and proportionally to the square of the scale ratio for the Mach scaling. The rate sensitivity increases proportionally to the square root

Table 5.4. **Roll attitude flying qualities parameters with and without stabilizer bar, and scaling effects based on Froude and Mach considerations. Percentage refers to fraction of full-stick deflection.**

		Stabilizer Bar		Scaling	
		with	without	Froude	Mach
Effective roll damping	$\frac{1}{\text{sec}}$	$-L_b(\tau_f + B_d\tau_s)$	$-L_b\tau_f$	$\sqrt{N}$	N
Control sensitivity	$\frac{\text{rad/sec}^2}{\%}$	$L_b(B_{lat} + B_d D_{lat})$	$L_b B_{lat}$	N	N^2
Rate sensitivity	$\frac{\text{rad/sec}}{\%}$	$\frac{B_{lat}+B_d D_{lat}}{\tau_f+B_d\tau_s}$	$\frac{B_{lat}}{\tau_f}$	$\sqrt{N}$	N
Rate resp. time const.	sec	$(L_b(\tau_f + B_d\tau_s))^{-1}$	$(L_b\tau_f)^{-1}$	$\frac{1}{\sqrt{N}}$	$\frac{1}{N}$

of the scale factor and proportionally to the scale ratio for the Froude and Mach model, respectively. And its rate response time constant decreases inversely to the square root of the scale factor (increase in bandwidth by a factor $\sqrt{N}$) and inversely to the scale ratio for the Froude and Mach models, respectively.

All of these effects contribute to making small-scale helicopters more agile and responsive, but also more difficult to control by the pilot. Note the even more dramatic effects for a Mach scale model.

Based on a Froude scaling, for a helicopter the size of the Yamaha R-50 ($N = 5$), the roll rate sensitivity is expected to be more than two times higher, and the roll rate time constant about two times smaller, than for the full-scale helicopter. Therefore, for good control, a pilot (or a control system), would need at least twice as much bandwidth as needed for a similar full-scale system. This requirement is difficult to satisfy for human pilots who are by nature bandwidth-limited. In the case of remotely controlled helicopters, the pilot also lacks the motion cues and visual cues that are crucial to the pilots of manned helicopters. It is specifically to help the pilot in his tasks, small-scale helicopters are usually equipped with stability augmentation systems, such as the stabilizer bar and an active yaw damping system. Next, we will look at numerical examples derived from the identification results.

Table 5.5. **Flying qualities parameters for R-50 and X-Cell based on identification results, with and without stabilizer bar. Also shown are expected ratios predicted by both Froude and Mach scaling considerations.**

		with/without Stab.		Ratio	Scaling Ratio	
		R-50	X-Cell	w/ Stab.	Froude	Mach
Effective roll damping	$\frac{1}{sec}$	-48/-8	-98/-17	2.0	1.4	2
Control sensitivity	$\frac{rad/sec^2}{\%}$	27.5/11.5	175/134	6.4	2	4
Rate sensitivity	$\frac{rad/sec}{\%}$	0.58/1.5	1.8/8	3.1	1.4	2
Rate resp. time const.	sec	0.02/0.13	0.01/0.06	0.48	0.7	0.5

R-50 and X-Cell Flying Qualities Comparison

Table 5.5 gives the numerical values for the attitude flying qualities metrics as predicted by the identified derivatives, along with the ratio of the X-Cell and the R-50 values. Also shown are the ratios predicted by both the Froude and Mach scaling. Notice the good agreement between the values for the rate sensitivity and the rate magnitude at 1 rad/sec shown in Figs. 5.3 and 5.4 (roll: 0.58 rad/sec = 33 deg/sec).

The stabilizer bar increases the effective roll damping and the effective control sensitivity. However, between these two parameters, the effective rotor damping increases most, resulting in a net decrease in the effective rate sensitivity. This effect of the stabilizer bar directly addresses the increase in effective rate sensitivity caused by the scaling.

Figure 5.8 illustrates the effect of Froude and Mach scaling on the attitude rate sensitivity. The figure shows the data points for the R-50 and X-Cell with and without stabilizer bar (normalized to the R-50 rate sensitivity). A Froude scale model will have an increased rate sensitivity. This effect is even more dramatic for the Mach scale model. Note that besides scale, the rotor time constant, the cyclic control authority, and the presence of a stabilizer (effective rotor time constant) have important effects on the rate sensitivity. Related through Mach scaling, we see that the X-Cell has a proportionally higher rate sensitivity, which is mostly attributed to a higher cyclic control authority.

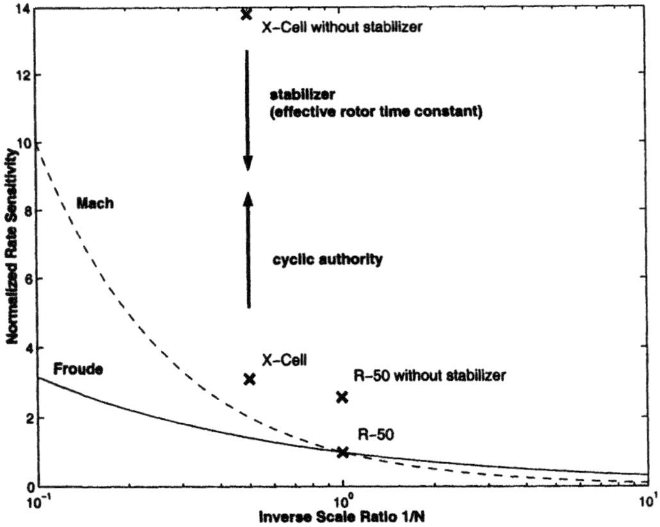

Figure 5.8. Effect of Froude and Mach scaling on the attitude rate sensitivity. Also shown are the effect of rotor stiffness and stabilizer bar.

To make the rate sensitivity scale independent, we would have to choose the rotor speed to be the same for the model as the prototype, i.e.

$$\tau_m = \frac{16}{\gamma \Omega_m} = \frac{16}{\gamma \Omega_p}. \tag{5.45}$$

Such a design constraint, however, would seriously affect the vehicle performance. Recall that the thrust-to-weight ratio curve for constant rotor speed shown in Fig. 5.6 decreases dramatically with decreasing vehicle size. Alternatively, we could also adjust the blade lock number to compensate for the effect of the rotor speed, i.e. for a Froude scale ($\Omega_m = \sqrt{N}\Omega_p$) we would need

$$\gamma_m = \frac{\gamma_p}{\sqrt{N}}, \tag{5.46}$$

which means that the lock number would have to be reduced (e.g. by increasing the blade moment of inertia) as the size of the vehicle is reduced. Such design requirements would also seriously impede other aspects of the vehicle flying qualities.

The stabilizer bar essentially allows one to influence the flying qualities (e.g. increasing the effective rotor time constant) without seriously affecting the

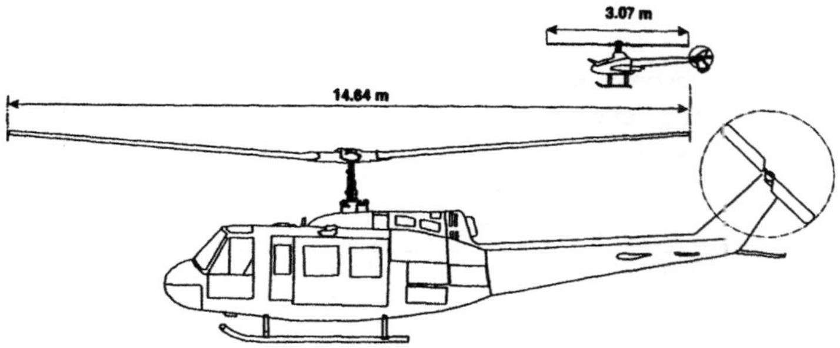

Figure 5.9. Bell UH-1H and Yamaha R-50 at their actual scale ratio

overall vehicle dynamics. The stabilizer bar provides additional degrees of freedom for the design of the flight characteristics.

4. Comparing Rotorcraft Through Scaling Rules

Now that we have meaningful scaling laws, we can relate differently sized rotorcraft by comparing their key configuration and dynamic parameters through scaling rules. We will first compare the key parameters of the identified R-50 to those of a full-size Bell UH-1H helicopter dynamically scaled to the R-50 size using the Froude scaling laws. The UH-1H was selected for the comparison because of its two-bladed teetering rotor; the UH-1H is also one of the few full-scale helicopters equipped with a stabilizer bar. Then we will compare the R-50 with the X-Cell helicopter using the Mach scaling rules.

4.1 Bell UH-1H vs. Yamaha R-50

Table 5.6 shows the comparison between the key configuration parameters and the identified dynamic characteristics for the R-50 with the model-scale equivalents for the UH-1H. We applied the set of Froude scaling rules (Table 5.2) for a scale ratio $N = 4.76$, or nearly $1/5^{th}$ scale. Figure 5.9 depicts both helicopters in their actual scale ratio.

Overall, the two helicopters are dynamically quite similar. Perfect similarity would require all scaled UH-1H parameters to be identical to the R-50's (i.e. 1 everywhere in the ratio column).

The largest dissimilarity occurs for the helicopter weight W and the blade moment of inertia I_β, where the R-50 shows values that are twice the values of the scaled UH-1H. The difference in weight is also responsible for a higher R-50

Table 5.6. Comparison between key configuration and dynamic parameters of the R-50 and a dynamically-scaled version of the full-scale UH-1H helicopter (scale ratio $N = 4.76$).

Parameters	UH-1H full-scale	Scaling rules	UH-1H scaled	R-50	Ratio
Rotor radius R (ft)	24	$1/N$	5.04	5.04	1
Rotor solidity σ	0.046	1	0.046	0.05	1.1
Weight W (lb)	8000	$1/N^3$	74	150	2
Rotor speed Ω (rad/sec)	34	$\sqrt{N}$	76.1	89.01	1.2
Blade inertia I_β (slug-ft^2)	1211	$1/N^5$	0.495	0.87	1.8
Blade Lock number γ	6.5	1	6.5	3.44	0.53
Normalized thrust coefficient C_T/σ	0.0606	1	0.0606	0.081	1.3
Normalized rotor hub height h/R	0.29	1	0.29	0.37	1.3
Flap spring L_b (rad/sec^2)	19.2	N	96.77	166.1	1.7
Natural frequency R-F mode $\omega_{\text{roll-flap}}$ (rad/sec)	4.38	$\sqrt{N}$	9.83	12.88	1.3
Non-dim. rotor flap time constant $\tau_s\Omega$ (rotor rev.)	5.7	1	5.7	30	5.3

normalized thrust coefficient C_T/σ, and the larger blade moment of inertia is responsible for a smaller R-50 blade lock number γ_b.

Another significant difference between parameters occurs for the flap spring L_b, for which the R-50 has a 60 % higher value. Since the flap spring derivative includes the effect of the fuselage moment of inertia, and since the R-50 has been shown to be comparatively heavier than the UH-1H, a comparison based on a similar fuselage moment of inertia would result in an even larger difference. This shows that the R-50 has a more rigid rotor head than that of a typical teetering head (which has no restraint). This is explained by the particular configuration of the R-50 rotor head; the R-50 has both a central teetering hinge and two flapping hinges (tri-hinge), and all hinges are fitted with elastomeric restraints.

However, by comparing the R-50 flapping spring derivative with that of the full-size BO 105 ($L_b = 523.5$; scaled for $N = 3.2$), we see that the R-50 rotor is not very rigid.

The natural frequency of the coupled R-F mode $\omega_{\text{roll-flap}}$ is directly related to the flap spring derivative (Eq. 5.24). The R-50 has a natural roll/flap frequency which is 20% higher than the scaled equivalent UH-1H frequency.

Finally, the R-50's non-dimensional rotor time constant (based on the stabilizer bar time constant; $\tau_s\Omega$) is more than five times larger than the UH-1H's. Note that without stabilizer bar both values would be significantly smaller (2.46 vs. 4.27 for the UH-1H and the R-50 respectively), which emphasizes the significance of the stabilizer bar on the R-50.

4.2 Yamaha R-50 vs. X-Cell

In the comparison between the X-Cell and the R-50, we used the Mach scaling rules to scale all of the X-Cell's key configuration and dynamic parameters up to the R-50 size. Table 5.7 shows the results that were obtained. Overall, the agreement between the R-50 values and the scaled X-Cell values through Mach scaling rules is good.

The configuration parameters, including the rotor solidity, the weight, the blade inertia scale according to our assumptions (geometrical similarity and constant density).

The rotor speed follows exactly the Mach scaling of Eq. 5.32. The biggest dissimilarity occurs for the normalized thrust coefficient, the blade lock number and flap spring derivative (natural frequency of the coupled R-F mode). It is somewhat surprising to see that perfect Mach scale model of the R-50 would have a greater rotor stiffness than that of the X-Cell.

Finally, the non-dimensional rotor time constants (based on the stabilizer bar time constant τ_s) are quite similar for both helicopters.

5. Further Scaling Considerations

In the following we will apply scaling considerations to other relevant aspects of small-scale rotorcraft performance and operation.

5.1 Scaling of the Speed Envelope

Unlike fixed-wing aircraft, rotorcraft operate from zero airspeed. Therefore, they see dramatic changes in their aerodynamic effects. The American Design

Table 5.7. **Comparison between key configuration and dynamic parameters of the R-50 and a dynamically-scaled version of the X-Cell helicopter (scale ratio $N = 0.5$).**

Parameters	X-Cell full-scale	Scaling rules	X-Cell scaled	R-50	Ratio
Rotor radius R (ft)	2.5	$1/N$	5.04	5.04	1
Rotor solidity σ	0.05	1	0.05	0.05	1
Weight W (lb)	17	$1/N^3$	136	150	1.1
Rotor speed Ω (rad/sec)	167	N	84	89	1.1
Blade inertia I_β (slug-ft^2)	0.026	$1/N^5$	0.83	0.87	1.1
Blade Lock number γ	2	1	2	3.8	1.9
Normalized thrust coefficient C_T/σ	0.042	1	0.042	0.081	2
Normalized rotor hub height h/R	0.31	1	0.31	0.37	1.2
Flap spring L_b (rad/sec^2)	320	N^2	80	166.1	2.1
Natural frequency rotor/fuselage mode $\omega_{\text{roll-flap}}$ (rad/sec)	17.9	N	8.95	12.88	1.4
Non-dim. rotor flap time constant $\tau_s\Omega$ (rotor rev.)	3.67	1	3.67	3.01	0.82

Standard (ADS-33 [2]) divides the rotorcraft speed envelope into the following three ranges:

- hover: 0 – 7.7 m/sec
- low speed: 7.7 – 23 m/sec
- forward flight: greater than 23 m/sec

Changes in airspeed affect the dynamics in different ways, including: changes in the relative airspeed at the blade; fuselage aerodynamic effects (fuselage drag and forces from the horizontal/vertical fin); and changes in the main and tail rotor working states. Hence, different scaling principles can be applied to relate the speed ranges of a full- and small-scale rotorcraft.

Advance Ratio Scaling

The nondimensional airspeed is given by the advance ratio μ . The advance ratio, defined as the ratio of the true airspeed V to the blade-tip speed (ΩR), is:

$$\mu = \frac{V \cos \alpha_D}{\Omega R}, \tag{5.47}$$

where α_D is the angle of attack between the free stream velocity and the rotor disc.

A meaningful way to scale the speed envelope is to preserve the same advance ratio. For a Froude-scale model, the rotor speed scales by the square root of the scale factor, and the rotor radius inversely to the scale factor. Thus, to preserve a same advance ratio, the airspeed needs to be scaled inversely to the square of the scale factor. The same conclusion can be reached using the units (ft/sec scales as $1/\sqrt{N}$). The equivalent speed ranges for the R-50 are therefore:

- hover: 0 – 3.4 m/sec
- low speed: 3.4 ?- 10.3 m/sec
- forward flight: greater than 10.3 m/sec

Note that the average airspeed attained during the cruise condition data collection for the R-50 was about 15 m/sec, which, based on this scaling, falls well into the forward flight speed range.

For a Mach scale model, the rotor tip speed stays unchanged, hence the speed envelope for the model will be the same as that of the prototype. This is in fact one of the reasons for using Mach scaled aircraft.

Inflow Ratio Scaling

An alternative approach is to scale the airspeed in terms of the rotor inflow velocity v_i. The motivation for this speed scaling is to reflect the aerodynamic working state of the rotor. As we will see next, the rotor inflow velocity scales inversely proportional to the square root of the scale ratio: $v_i \propto 1/\sqrt{N}$, hence, to keep the ratio V/v_i the same for the model and prototype, the speed envelope must be scaled by $1\sqrt{N}$, which leads to the same speed ranges as obtained for the Froude scaling.

5.2 Rotor Performance and Scaling

Because gross vehicle weight decreases faster than rotor disc surface area, another effect of scale is a decrease in the rotor disc loading ($D.L. = W/A \propto$

Table 5.8. **Rotor aerodynamic parameters for different helicopters near hover.**

	UH-1H	R22	R-50	CMU's	X-Cell	MIT's
N	0.21	0.38	1	1	2	2
D.L. (lbf/ft^2)	4.4	2.45	1.24	1.9	0.51	0.87
v_i (ft/sec)	30	23	16	20	10	14
λ_i	0.037	0.033	0.036	0.045	0.025	0.032
τ_λ	0.17	0.12	0.066	0.053	0.051	0.040
Re $_{\text{tip}}$	9.0×10^6	2.6×10^5	9.9×10^5	9.9×10^5	5.2×10^5	5.2×10^5
$C_{T,\text{hover}}$	0.00279	0.00217	0.00262	0.00405	0.00122	0.00208

$1/N$) and also in the induced inflow velocity: ($v_i \propto 1/\sqrt{N}$) from

$$v_i = \sqrt{\frac{D.L.}{2\rho}}. \tag{5.48}$$

A smaller induced velocity means a smaller induced power ($P_i = Tv_i$), the principal power loss in hovering flight. The rotor of a small-scale rotorcraft should therefore be $\sqrt{N}$ times more efficient than its prototype when measured per pound of lift produced. Recall that the induced inflow ratio λ_i (ratio of induced velocity to blade tip speed) for a rotor in hover is given by:

$$\lambda_i = \frac{v_i}{\Omega R}. \tag{5.49}$$

As a nondimensional quantity, this ratio should remain relatively unchanged by scale. This fact can be observed from the numerical examples in Table 5.8.

An important parameter for the transient response of a rotorcraft in both vertical flight and attitude response is the inflow settling time. The inflow settling time can be approximated by (see [68]):

$$\tau_\lambda = \frac{0.849}{4\lambda_i\Omega}. \tag{5.50}$$

Reading from Table 5.2, we note that τ_λ should decrease as $1/N$ for Froude models and as $1/\sqrt{N}$ for Mach models. This suggests that rotor inflow effects are less important in small-scale rotorcraft.

Table 5.8 lists these and other rotor aerodynamic quantities for the four vehicles in question (CMU's and MIT's denote the instrumented vehicles). Overall,

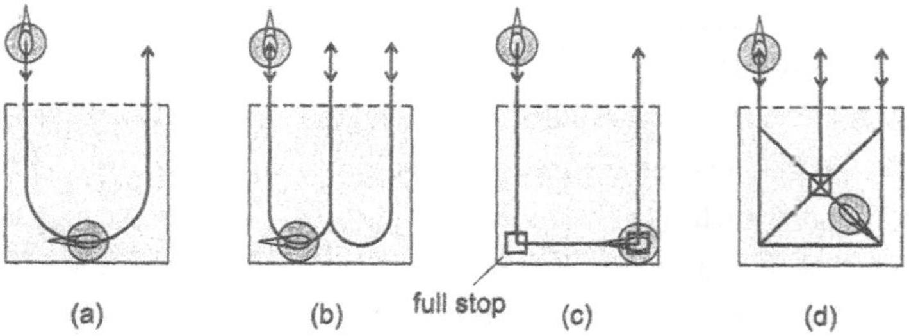

Figure 5.10. **Illustration of different flight patterns that can be used to explore a 2-dimensional confined space. (a) and (b) show fixed-wing type of flight patterns, and (c)–(d) show patterns with loitering.**

the variations between these parameters are consistent with our scaling hypotheses.

5.3 Maneuvering and Flight Operations

To illustrate the combined effect of scaling and design on maneuvering and flight operations, we can look at a simple coordinated level-flight turn maneuver. For a given cruise speed, the minimum turn radius is determined by the maximum load factor. We can obtain an approximation for the maximum load factor $\bar{L}$ from the identified maximum vertical acceleration, which was shown in Figure 5.7. We obtain approximations for the load factors of $\bar{L} \approx 1.1g$ for the R-50 and $\bar{L} \approx 2.1g$ for the X-Cell. These values correspond to maximum bank angles $\bar{\phi}$ of about 27 and 62 degrees for the R-50 and X-Cell, respectively. The minimum turn radius is then approximated by

$$\rho_{\min} \approx \frac{v^2}{\bar{L}\sin\bar{\phi}}. \tag{5.51}$$

Evaluating this expression for 5 and 15 m/sec, we obtain minimum turn radii $\rho_{\min}$ = of 1.4 and 12 m for the X-Cell, and 5 and 45 m for the R-50. The R-50 would therefore take about four times more space than the X-Cell to execute a 180-degree turn, which at faster speeds can mean a serious operational handicap.

The maximum rectilinear acceleration can also be estimated from these values. Specifically, $\bar{a}_x \approx \bar{L}\sin\bar{\phi} = 1.89g$ and $0.5g$ for the X-Cell and R-50, respectively. In terms of rectilinear motion, the X-Cell exceeds the gravitational acceleration in all three dimensions, making it virtually possible to cancel the effect of gravity. When combined with high angular sensitivity, quick redirections of the thrust vector and thus rapid changes in the flight direction can be achieved by the X-Cell.

Such difference in maneuverability can have an important impact on operational agility [67]. Consider for example the problem of exploring a confined 2-dimensional space such as illustrated in Fig. 5.10. The speed and accuracy at which such a task can be accomplished will be a direct function of the maneuverability of the vehicle. For example, (a) refers to flying without stoping; the maximum airspeed will be determined by the maximum load factor and the size of the space. (b) shows multiple runs with a median track, and (c)–(d) include hovering points.

6. Stabilizer Bar Effects

In the following, we will describe how the stabilizer bar acts as a control augmentation and therefore influences the stability and transient response characteristics of the vehicle. We will illustrate its effects through simulations with our identified model; we will compare the helicopter's response to pilot commands and gusts for the cases with and without a stabilizer bar.

For a plain rotor, the low-frequency attitude rate response magnitude decreases at these low frequencies due to the effect of the airspeed on the rotor flapping and thus on the attitude dynamics, i.e. the long-term attitude response following a pilot cyclic input is "washed-out" by the airspeed effects. This effect is less severe for a rotor equipped with a stabilizer bar, as will be explained later in this section.

Since the stabilizer bar acts as a control augmentation it partially masks the vehicle dynamics (recall Eq. 2.28). Therefore, when looking at the model without the stabilizer bar equations, we must keep in mind that the low-frequency characteristics may be inaccurate (below about 1 rad/sec).

6.1 Coupled Rotor Stabilizer Equations

If we look at the transfer function for the the stabilizer bar flapping response derived in Eq. 5.12 we can see that it will exhibit a response to both vehicle angular rate and control input. By taking into account the coupling mechanism between the stabilizer bar and the rotor (see Eq. 3.91), we can represent the coupled fuselage-rotor-stabilizer system as a feedback system, where the stabilizer bar takes on the role of a control augmentation (see Fig. 5.11).

Based on this representation we see that the stabilizer bar acts first as a lagged rate feedback (or pseudo attitude feedback), which corresponds to the

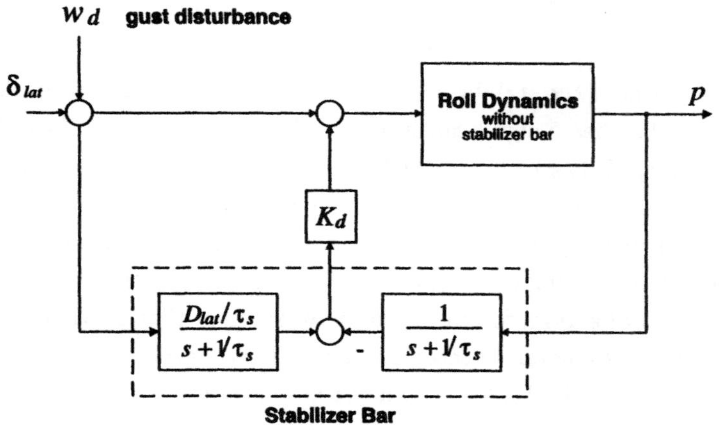

Figure 5.11. **Block diagram showing the stabilizer as a control augmentation. Note how the stabilizer bar acts as a lagged rate feedback as well as a cyclic command booster.**

response of the stabilizer to the helicopter roll rate p; and second, as a lagged control booster, which corresponds to the response of the stabilizer to the pilot input. Both actions are achieved by augmenting the main rotor cyclic input (δ_{lat}) via the Bell mixer, which is modeled as a gain K_d.

Next, we will illustrate the effect of the stabilizer bar from simulation results. First, we will give a physical interpretation of the stabilizer bar.

6.2 Physical Interpretation of the Stabilizer Bar

The main effect of the lagged rate feedback action of the stabilizer bar is to increase the damping moment in the helicopter attitude dynamics (see Eq. 5.17), changing how the helicopter responds to the pilot commands and disturbances.

The main rotor provides for most of the attitude dynamics' natural damping. From Eq. 5.4, we see that the key parameter of the rotor damping is the rotor time constant τ_f). The rotor time constant determines how fast the rotor will flap following an excitation such as a cyclic command or wind gust. The rotor also responds to the helicopter roll and pitch motion. For a step-like roll rate, the rotor time constant determines how much the rotor would lag behind the rotor shaft. For example, for a helicopter roll rate p, the steady-state rotor lag angle would be $\tau_f p$. By lagging behind the helicopter shaft, the rotor produces a moment on the helicopter in a direction opposite, and of a magnitude proportional, to the roll rate p. The total damping moment produced by the rotor is:

$$M_{damp} = -L_b \tau_f p \tag{5.52}$$

The main effect of the stabilizer bar is to increase the natural rotor damping. The stabilizer bar has a significantly larger time constant than does the rotor (for the R-50: τ_s=0.35 sec vs. τ_f=0.046 sec); therefore, it has a slower response and would lag behind both the main rotor tip-path-plane and the rotor shaft. Since the main rotor cyclic input is augmented by a component proportional to the stabilizer bar flapping $\delta_{aug,stab} = B_{lat}K_d d$, the lagging of the stabilizer bar, through its control action, forces the rotor to lag more than it would naturally, which increases the amount of rotor damping. This effect is visible in the flying qualities summarized in Table 5.4 and evaluated in Table 5.5.

A drawback of the stabilizer bar is that while it increases the natural rotor damping, it decreases the damping of the coupled R-F mode. This happens because the stabilizer bar increases the effective rotor time constant, which then has a characteristic response time that is closer to the natural frequency of the coupled R-F mode (recall Eqs. 5.24 and 5.25). The former effect is usually perceived as an improvement in the handling qualities, for example it permits a human pilot to control small-scale helicopters. The latter effect, however, increases the tendency for pilot-induced oscillation (PIO). For a small-scale helicopter, the natural frequency of the coupled R-F mode is usually beyond the piloted excitations. However, as we will see in next chapter, this lightly damped R-F model can be an issue for the implementation of high-bandwidth control systems.

6.3 Simulation of the Stabilizer Bar

To further illustrate the effects of the stabilizer bar, we simulated the response of the system shown in Figure 5.11 to a lateral pilot command and to a lateral input disturbance for the case with and without a stabilizer bar. For the simulation, we used the identified R-50 roll dynamics.

To avoid exciting the coupled R-F mode, we filtered the step for both the pilot command and input disturbance using a specially designed notch filter. A time history of the input is shown in Figure 5.12. The details of the notch filter will be presented in next chapter.

Response to Gust Disturbance

A wind gust disturbs the direction and speed of the airflow encountered by the helicopter. This changes the aerodynamic load of the rotor, producing an

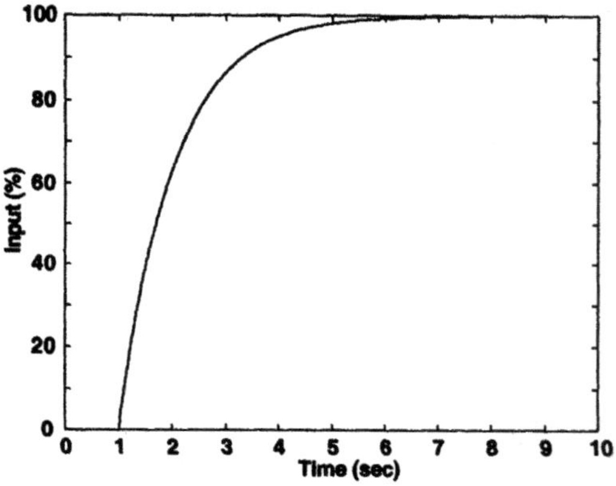

Figure 5.12. Time history of the input used for the pilot command and for the input disturbance. We used a filtered step to avoid exciting the lightly damped fuselage-rotor-stabilizer mode.

uncommanded rotor flapping response which will, in turn, disturb the helicopter attitude. If we ignore the forces produced by the gust on the fuselage, a wind gust, since it changes the aerodynamic angle of attack at the blade (in a cyclic manner), has an effect on the helicopter similar to that of a sudden change in the cyclic control δ_{lat}. Thus wind gusts are effectively modeled as input disturbances (shown as w_d in Figure 5.11).

The result of the simulation is shown in Figure 5.13. As can be observed, the response with the stabilizer bar is significantly smaller and has a slower development than the one without the stabilizer bar.

Related to the gust response is the response to airspeed when going from hover to forward flight. In cruise flight, the speed asymmetry causes the rotor to flap back, and the pilot uses a significant amount of the longitudinal cyclic control authority to trim the aircraft. The stabilizer bar, with its small Lock number, is less subject to aerodynamic forces. Thus it will have a weaker tendency to respond to airspeed variations and will also play a role in trimming the vehicle. The reduced sensitivity to airspeed will also impact the translational dynamics of the aircraft, in particular the important phugoid mode, which will be explained later in this chapter.

Response to Pilot Commands

Since the pilot command is applied at the same location as the input disturbance, and since the same input time history is used, the helicopter response

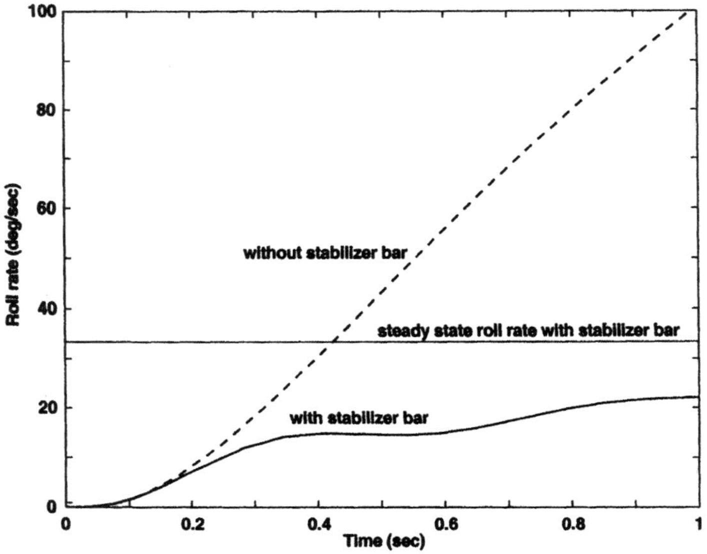

Figure 5.13. Comparison between the roll rate response to a cyclic input for the case with (solid) and without (dashed) the stabilizer bar. Note that the input can be a pilot command or a gust disturbance.

to the pilot's command is the same as the one presented for the gust rejection (Figure 5.13). Notice that the steady-state roll rate corresponds to the rate sensitivity given in Table 5.5 (0.58 rad/sec = 33 deg/sec).

The difference between the two systems is also demonstrated by the roll rate frequency response depicted in Figure 5.14. We see that, for the case without a stabilizer bar, the magnitude is about 15 dB higher than the magnitude with a stabilizer bar. The system with a stabilizer bar also exhibits the effect of lightly damped R-F mode.

6.4 Vehicle Stability

To determine how the stabilizer bar affects the vehicle stability characteristics, we proceeded with a root-locus analysis. As for the previous analysis, we concentrated on the lateral dynamics.

Figure 5.15 shows how the poles and zeros of the R-50's lateral dynamics change as a function of the stabilizer bar gearing K_d (gain range: 0 to 30 %/rad). The poles shown in this Figure are associated with the following natural modes of motion:

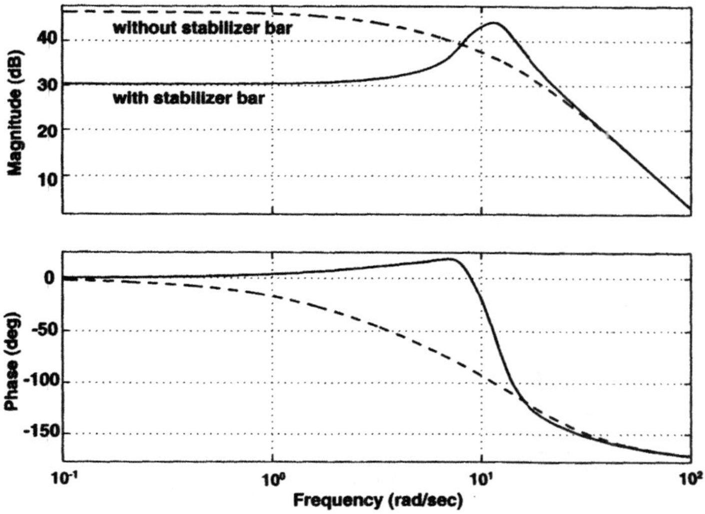

Figure 5.14. **Comparison between the frequency response from lateral input to roll rate for the case with (solid) and without (dashed) the stabilizer bar.**

(a) complex conjugate eigenvalues of the coupled R-F mode. This is an oscillatory roll mode involving a strong coupling with an oscillatory rotor/stabilizer motion. By increasing the stabilizer gearing, we see that the eigenvalue pair goes toward the right-half plane; approaching the imaginary axis, the damping of the R-F mode decreases, and the natural frequency of the oscillations increases. The decrease in damping of that mode is visible in Fig. 5.16.

(b) stable roll mode. Increasing K_d reduces the mode's time constant.

(c) stable roll-velocity mode. Increasing K_d increases the mode's time constant.

(d) unstable roll-velocity mode. Increasing K_d decreases the divergence of the unstable mode.

7. Modal Characteristics of the R-50

The response of a linear system to any type of input can be described as a linear combination of its natural motions or dynamic modes. The dynamic modes provide important information about the system's dynamic characteristics, such as its stability and the type of motion that can be encountered. The

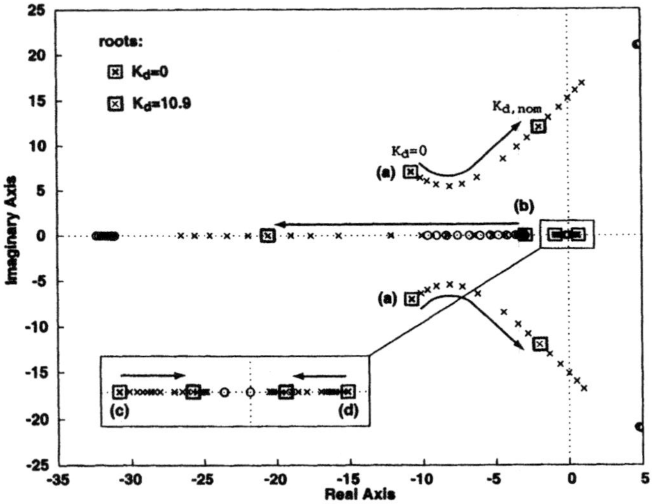

Figure 5.15. Root locus showing the effect of different stabilizer gain K_d (0 to 30 %/rad) on the roll dynamics (nominal value shown as $K_{d,\text{nom}} = 10.9$). The eigenvalues correspond to the following modes: (a) coupled fuselage-rotor-stabilizer mode; (b) roll mode; (c) stable roll-velocity mode; (d) unstable roll-velocity mode.

modes of a linear system are found by examining the system's eigenvalues and eigenvectors.

Our 13th order system (see state-space Eq. 3.110) has $n = 13$ eigenvalues λ_n and eigenvectors x_n. These are calculated from the $n \times n$ system matrix F by solving the eigenvalue problem [9]:

$$\det[\lambda I - F] = 0 \tag{5.53}$$

The dynamics of helicopters are highly coupled, therefore most of the modes are coupled. The typical helicopter modes are (see [68]):

- long-period fuselage-motion mode characterized by speed excursions and an exchange between kinetic and potential energy (phugoid mode)
- short-period lightly damped mode characterized by attitude excursion coupled with rotor flapping excursions (coupled R-F mode)
- long-period fuselage roll-motion with lateral velocity and yaw motion (Dutch roll)
- various aperiodic motions, such as the roll subsidence mode, the spiral subsidence mode, or the heave subsidence mode

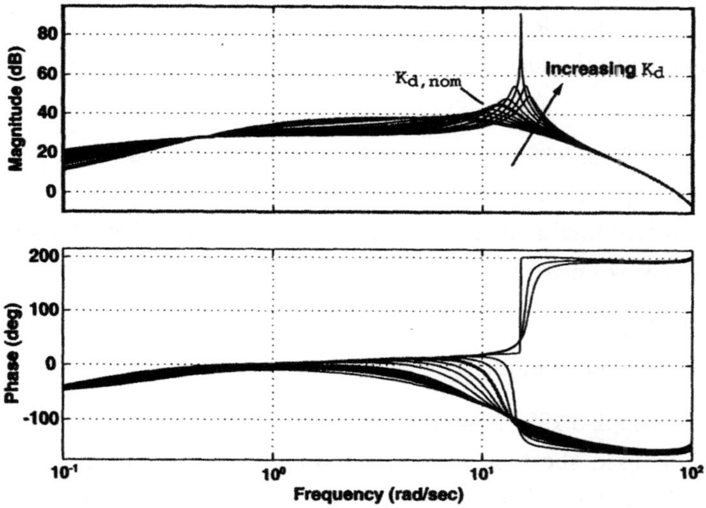

Figure 5.16. Frequency response of the roll rate as a function of the stabilizer bar gain K_d. Notice the decrease in damping of the R-F mode with increase in K_d. We can also see how increasing K_d decreases the airspeed's "washout effect" on the low frequency rate tracking, i.e. it reduces the sensitivity of the attitude dynamics to airspeed.

The exact composition of the modes changes from one aircraft to another. The analysis and interpretation of a helicopter's modal characteristics can be difficult. For the interpretation of the R-50 modes, the complex eigenvectors were normalized and represented graphically in polar coordinates. This way, we can see how the components that compose each mode relate in magnitude and in phase. Note that only the components with the highest magnitude are included in the graphs. Also, the modes from the previous section will not come out exactly as they did previously because of the coupling between the other dynamics (longitudinal, vertical and yaw).

Figures 5.17 and 5.18 show the graphs for, respectively, the hover and the cruise flight conditions. Note that, due to the limited low-frequency model accuracy (< 0.32 rad/sec and < 0.64 rad/sec for, respectively, the hover and cruise flight conditions), the slow modes such as the phugoid modes may lack accuracy.

Eigenvalues and Modes in Hover Flight

In hover flight, the eigenvectors and modes are the following:

- $\lambda_{1,2} = 0.3061 \pm 0.094i$ ($\zeta = -0.9562$; $\omega = 0.3201$ rad/sec)
  Unstable, damped, long period oscillatory, phugoid type mode. The main components are the longitudinal and lateral velocities and the roll and pitch angles

- $\lambda_{3,4} = -0.4007 \pm 0.086i$ ($\zeta = 0.9778$; $\omega = 0.4098$ rad/sec)
  Damped, long period oscillatory, phugoid type mode. The main components are the longitudinal and lateral velocities, the vertical velocity, and the roll, pitch, and yaw angles

- $\lambda_5 = -0.6079$
  Damped yaw-heave mode

- $\lambda_{6,7} = -1.699 \pm 8.192i$ ($\zeta = 0.2031$; $\omega = 8.366$ rad/sec)
  Lightly damped short period pitch mode. The main components are the pitch and roll rate. The mode has a 25% roll rate component. This mode is related to the coupled fuselage-rotor-stabilizer bar dynamics

- $\lambda_{8,9} = -6.196 \pm 8.198i$ ($\zeta = 0.6029$; $\omega = 10.28$ rad/sec)
  Under-damped short period yaw mode (effect of active yaw damping)

- $\lambda_{10,11} = -2.662 \pm 11.58i$ ($\zeta = 0.2241$; $\omega = 11.88$ rad/sec)
  Lightly damped short period roll mode. The main components are the roll rate, the yaw rate, and the pitch rate. This mode is related to the coupled fuselage-rotor-stabilizer bar dynamics

- $\lambda_{12,13} = -20.17 \pm 4.696i$ ($\zeta = 0.9739$; $\omega = 20.71$ rad/sec)
  Nearly critically damped short period oscillatory roll mode. The main components are the roll and pitch (40%) rate. This mode is related to the coupled fuselage-rotor-stabilizer bar dynamics

Eigenvalues and Modes in Cruise Flight

From the cruise flight model, we obtain the following eigenvalues:

- $\lambda_1 = -0.1216$
  Damped longitudinal mode

- $\lambda_2 = -0.9614$
  Damped yaw-heave mode with a component of roll motion

- $\lambda_3 = -1.838$
  Damped yaw mode with a lateral velocity component (tail-fin effect)

- $\lambda_{4,5} = -2.321 \pm 8.794i$ ($\zeta = 0.2552$; $\omega = 9.095$ rad/sec)
  Lightly damped short period pitching mode with a large (90%) rolling component. This mode is related to the coupled fuselage-rotor-stabilizer bar dynamics

- $\lambda_{6,7} = -5.005 \pm 8.133i$ ($\zeta = 0.5241$; $\omega = 9.549$ rad/sec)
  Under-damped short period yawing mode (active yaw damping effect)

- $\lambda_{8,9} = -3.396 \pm 12.43i$ ($\zeta = 0.2636$; $\omega = 12.88$ rad/sec)
  Lightly damped short period rolling mode with a small (10%) pitching component. This mode is related to the coupled fuselage-rotor-stabilizer bar dynamics

- $\lambda_{10,11} = -27.04 \pm 7.019i$ ($\zeta = 0.9679$; $\omega = 27.94$ rad/sec)
  Nearly critically damped short period rolling mode with a 40% pitching component

Note that in cruise flight the hover phugoid modes ($\lambda_{1,2}$ and $\lambda_{3,4}$) are absent. This is due to the speed derivatives ($L_{u,v}$ and $M_{u,v}$) which had to be set to zero during the cruise condition identification. Instead, they are replaced by a damped longitudinal mode (λ_1) and a damped lateral-directional mode (λ_3). The two remaining eigenvalues, which are not shown in the list above, are located at the origin.

8. Conclusions Regarding Small-Scale Helicopter Dynamics

By applying the Froude and Mach scaling rules to the key parameters of our model, we were able to determine the effect of scale on the rotorcraft dynamics. The scaling effects confirmed that reducing the size of a rotorcraft causes its sensitivity to control inputs and, at the same time, its bandwidth, to increase. This translates into an increase in agility but, at the same time, increases the demands on the pilot to successfully control the vehicle. We showed that the stabilizer bar improves the pilotability (at the cost of reducing the damping of the coupled R-F mode) by acting as a lagged rate feedback, which increases the effective roll damping and thus decreases the rate sensitivity and bandwidth.

Through this analysis, we were also able to show that the R-50 was dynamically quite similar to the full-size Bell UH-1H. The R-50 is proportionally heavier (aircraft weight and blade moment of inertia) and has a larger rotor stiffness due to the elastomeric restraints and the teetering/flapping hinge com-

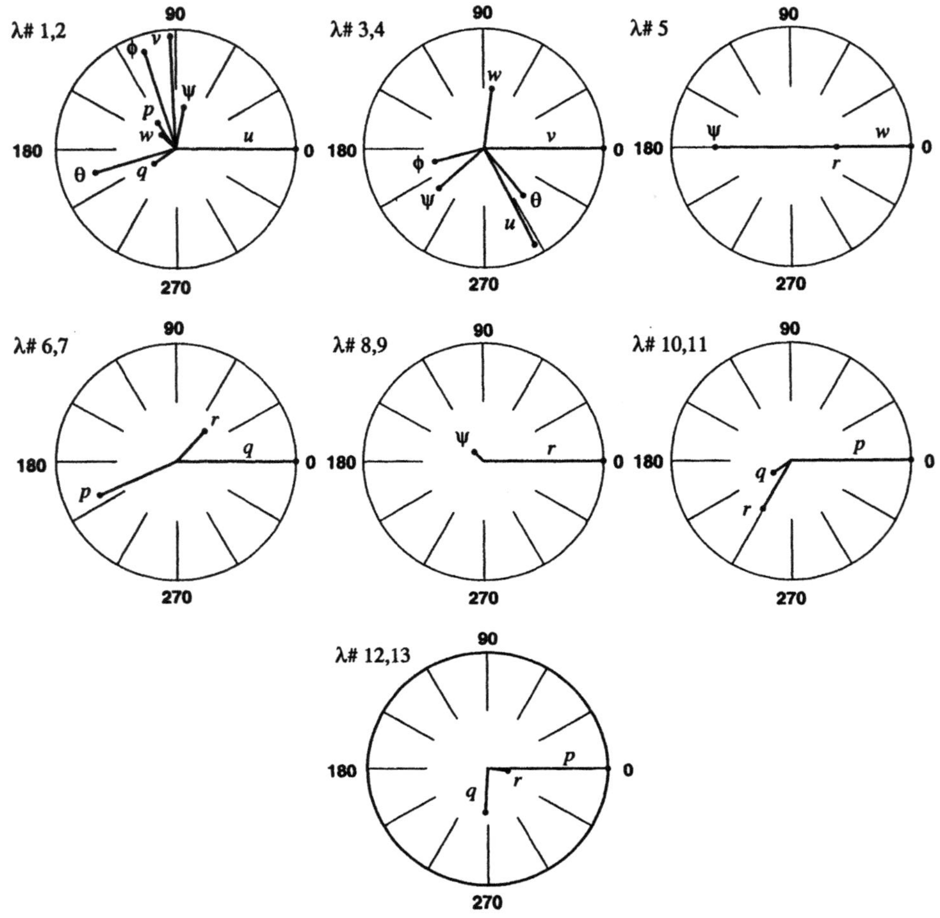

Figure 5.17. **Graphical representation of the eigenvectors derived from the identified hover model. Each eigenvector is represented in a phase diagram showing its dominant components.**

bination. We were also able to show that the dynamics of both helicopters are strongly influenced by the stabilizer bar. At the same time, the scaling rules showed that the X-Cell does not belong to this class of dynamically similar helicopter, but relates best to the R-50 through Mach scaling rules. Vehicles belonging to this class are characterized by unusually high thrust-to-weight ratio and fast attitude dynamics.

The application of the scaling rules has limitations. In particular, it assumes that the mechanical properties of the rotor head and other design features

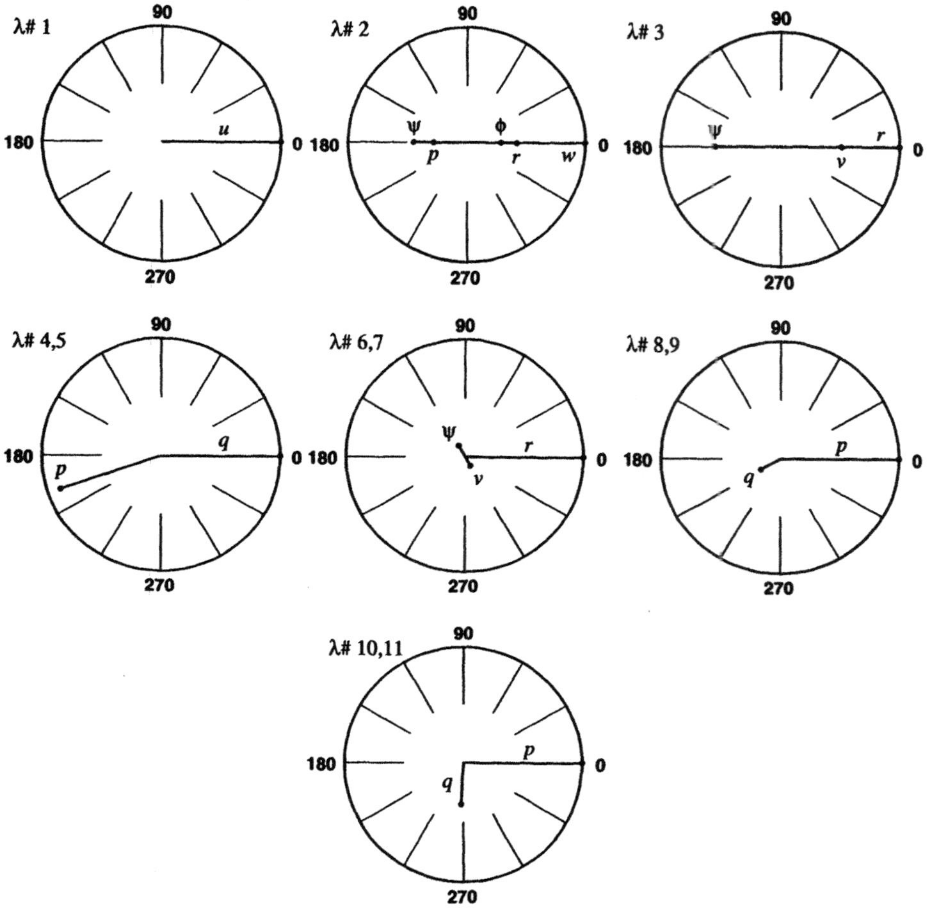

Figure 5.18. **Graphical representation of the eigenvectors derived from the identified cruise model. Each eigenvector is represented in a phase diagram showing its dominant components.**

do not change with scale. However, the scaling rules do give us an idea of how key physical flying qualities parameters can be expected to change with scale. These are important results that can be used when comparing different helicopters and determining scale-adequate control performance specifications.

In the next chapter, we will see how some of the characteristics influence the control design task.

Chapter 6

ELEMENTS OF CONTROL DESIGN

An accurate vehicle model like our identified linear model opens the door for a range of model-based control design methodologies. The control design method has an influence on the performance that can be attained. At the same time, it defines how the design problem is posed, what information is needed from the system, and what tools (and knowledge) are necessary to solve it. For example, a robust H_∞ control approach will likely yield the best performance a linear controller can achieve. However, to do so, it will require a highly accurate model of the system. On the other end, a classical controller can be setup very quickly, without the need for a model. For a complex system like a rotorcraft, it will yield performance significantly inferior to that of a competing H_∞ design. However, with some better understanding of the vehicle dynamics the potential of a classical architecture can be significantly enhanced.

Classical methodologies based on PID (Proportional-Internal-Derivative) feedback are still prevalent in the practice of full-scale rotorcraft control. Robust control design methods have been applied but are not yet widely used (see the line of work of Postlethwaite [96, 90, 78] with H_∞). For unmanned rotorcraft, both classic and robust control methods have been applied, but limited flight experience exists with the modern methods due to the modeling shortcoming. Recent milestones attained with small-scale rotorcraft control using modern techniques will be outlined in the next chapter.

The choice of synthesis or design methodology is as much a function of the desired performance as it is a function of other factors, such as the available resources (e.g. design tools, engineer training) and implementation constraints (e.g. sensing, design philosophy, certification requirements). In the following will not try to propose a particular approach. Instead, the goal is to introduce the control problem from a broad physical perspective, presenting elements of analysis, design, optimization, and validation. We will keep the scope mostly to the vehicle and thus not go very deep into the control formalism (an excellent multivariable control perspective is available in Weilenmann's thesis [92]). We will try to relate the control questions and problems to the material presented earlier. Furthermore, as we will see in this chapter, system identification also plays a fundamental role in the analysis, design, and implementation of the control system.

This chapter is presented in the form of a case study, where we use our identified vehicle model and closed-loop system identification techniques to analyze the classic feedback control system originally used on Carnegie Mellon's Yamaha R-50. Such a classical architecture is well suited to this task, as it provides an intuitive understanding of the mechanisms involved in rotorcraft control. The analysis will actually allow us to identify a performance limitation that has plagued classical controllers in small-scale rotorcraft. As a design example, we develop a solution that addresses this limitation and allows significantly higher performance to be attained with a classic control architectures. We will use frequency domain identification for the validation and evaluation of the enhanced closed-loop system. Following the analysis, we optimize the performance of the enhanced PD control system using CONDUIT (CONtrol Design Unified InTerface [22]), to illustrate the gain in achievable performance. Finally, we introduce the notions of flying qualities by giving a brief overview of the basic criteria used for manned rotorcraft and applying them to our optimized controller. This will provide some insight into the problem of selecting and defining criteria which could potentially be used as performance specifications for the design of automatic control systems for unmanned vehicles.

1. Classical Rotorcraft Control System

The simplest control systems used in rotorcraft are based on single-input single-output (SISO) proportional-integral-derivative (PID) feedback loops. Such a system was originally used on Carnegie Mellon's (CMU) RUAV and provided sufficient performance to accomplish basic position and velocity tracking near hover flight [63, 3]. The big advantage of this simple feedback architecture is that it can be implemented without a model of the vehicle dynamics; all feedback gains can be turned on empirically in flight. However, as we will see in this chapter, the performance that can be attained with such a controller is rapidly reaches limits without a more detailed analysis.

1.1 Description of the PD Control System

General Rotorcraft Control Principle

Longitudinal and lateral velocity control is performed through the vehicle pitch and roll dynamics. This is due to the necessary tilting of the helicopter in order for the rotor to produce sufficiently large horizontal forces (propulsive thrust components). The larger the pitch or roll angle the bigger the longitudinal, respectively lateral, propulsive force. Hence, in steady state, the helicopter velocity will be proportional to its attitude.

To implement this control principle, a pitch and roll attitude control system is often subordinated to a, respectively, longitudinal and lateral velocity control system. Such a control architecture is called nested or cascaded. To work, however, the inner attitude control loop must have a higher bandwidth than the outer velocity control loop. This is usually not a problem in helicopters where the rotational dynamics are naturally faster than the translational dynamics. The attitude and position, or velocity loops, can then be designed sequentially, which simplifies the control design task. For high-performance helicopters such as the X-Cell, or helicopters where this bandwidth separation is not sufficient, a simultaneous design is necessary.

CMU's Position Control System

CMU's PID controller is implemented as aposition command system. A block-diagram is shown in Figure 6.1. The position control system works on the same principle as the velocity control system. In this system, the attitude controller is composed of simple attitude angle feedbacks (with gains K_θ and K_ϕ). These control loops track the reference pitch θ_{ref} and roll ϕ_{ref} attitude angles which are produced by the longitudinal and lateral position control loops, respectively. These position control loops have both proportional and derivative feedback. For example, for the longitudinal position, a proportional gain K_x is applied to the longitudinal position error $e_x = x - x_{ref}$, and a derivative gain K_u is applied to the estimated longitudinal body velocity u (the position references x_{ref} and y_{ref} are in body-fixed coordinates).

This system was originally designed for position hold and small displacements. To track trajectories with such a system, trajectories must be broken down in a sequence of waypoints that are feasible for the control system. Tra-

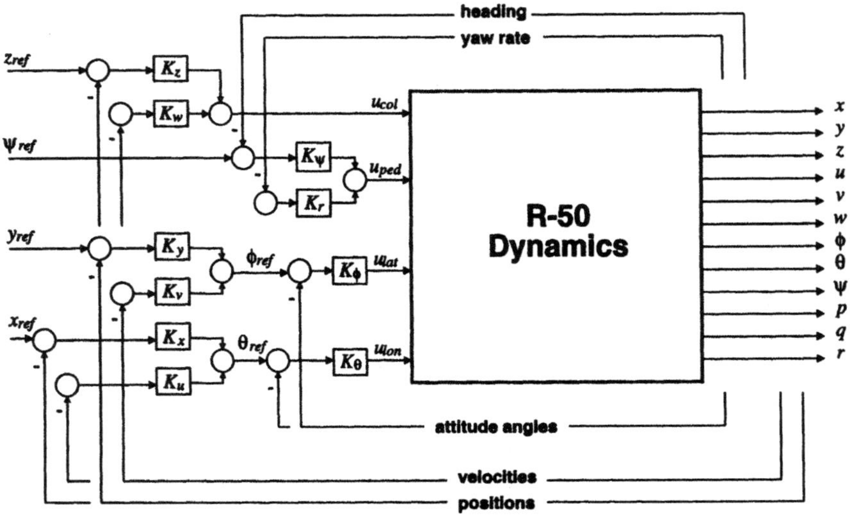

Figure 6.1. Block-diagram of the baseline PD control system. Note the two nested attitude control loops, with feedback gains K_ϕ and K_θ. Sensor dynamics are ignored.

jectory tracking would be easier to perform with a velocity control system. The vertical position z and heading ψ controllers use proportional and derivative feedback loops. Notice that all feedback loops in this position control system are SISO.

1.2 Simulation of the Position Controller

Figure 6.2 shows the comparison between the actual and predicted vehicle responses for a step-like position command applied successively to the lateral and longitudinal axis. For the comparison, we applied the same 4-meter reference step to the model of the closed-loop system. The trim values at the instant of the application of the step were subtracted from the recorded responses. The results show good agreement for the on-axis responses; a small time shift that can be attributed to some unmodeled delay is noticeable in the roll and pitch responses. This delay will be identified later using closed-loop identification. The good closed-loop agreement demonstrates that the model can be used for the optimization of the closed-loop performance.

Figure 6.2 is also a good illustration of the helicopter control modality. The lateral and longitudinal motion is achieved via a change in the vehicle's roll and pitch attitude. We notice that the baseline PD control system exhibits

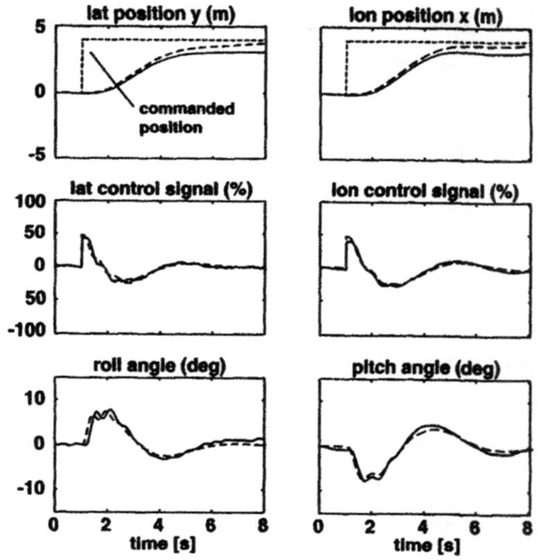

Figure 6.2. Comparison between model-predicted (dashed) and real (solid) closed-loop responses for a 4-meter step-like position command in the lateral direction (left) and longitudinal direction (right).

some attitude oscillations. These are due to the lightly damped coupled rotor-fuselage (R-F) mode. Such oscillations can be objectionable for certain types of operations and also show that the system is operating at the limits of its stability. In the following, we will analyze the stability and performance of the controller using results obtained from closed-loop frequency domain system identification, focusing on the attitude controller.

2. Analysis of the Attitude Controller

To provide a detailed analysis of the performance of the PD control system, we proceed with a closed-loop identification. Using the same technique as for the open-loop dynamics described in Chapter 2, we can estimate the frequency response of the closed-loop system. These frequency responses correspond to the transfer functions of the closed-loop vehicle-controller system.

We can determine the stability of the closed-loop system by examination of the phase and gain margin (Bode Analysis). Phase and gain margins can be readily determined from the model of the closed-loop system. However, to further validate the model in a closed-loop setting, we will compare the stability margins predicted by the model with those derived from the closed-loop identification. In the following, we will show how we can extract the loop-gain functions used in the Bode analysis from the identified closed-loop frequency responses. We concentrate on the analysis of the roll and pitch attitude controllers in hover flight condition. The approach is similar for the other control axes.

2.1 Closed-Loop System Identification

For the closed-loop identification we proceed similarly to the open-loop identification of the airframe dynamics. We collect flight data using frequency sweep techniques. In the present case, the inputs are the attitude pitch and roll attitude commands. The outputs that are of direct interest for the present analysis are the attitude angles and rates. From the collected flight data, we compute the frequency responses between the commanded attitudes and the resulting aircraft attitude using the frequency domain analysis tools provided in CIFER. The identified closed-loop frequency responses are used to validate the model of the closed-loop system and extract the loop gain functions that are used for the performance and stability analysis.

Data Collection Technique

Similar to open-loop identification, the accuracy of the frequency response estimates will depend on the quality and quantity of the data collected; the same experimental guidelines basically apply. The closed-loop roll and pitch examples shown herein are based on two frequency sweeps applied to each control axis. As a result, the identified frequency responses are noisy. The lack of data is most pronounced in the range of frequencies of the coupled R-F mode. There, the magnitude of the sweeps had to be deliberately reduced due to sensitive equipment carried by the RUAV at the time of the experiment. The low-frequency range is generally affected by a lack of data points; this was already the case for the identification of the airframe dynamics. The quality of the frequency responses is nonetheless sufficient to validate the accuracy of the model.

Identification of the Attitude Controller

From the time responses collected during the closed-loop experiments we obtained estimates for the frequency responses characterizing the response of the helicopter under the attitude control system. Four frequency responses were estimated:

- the on-axes pitch $H_{\theta,\theta_{ref}}$ and roll response $H_{\phi,\phi_{ref}}$,
- and the off-axes responses: $H_{\theta,\phi_{ref}}$ and $H_{\phi,\theta_{ref}}$,

where ϕ_{ref} and θ_{ref} are the roll and pitch attitude reference commands.

These frequency responses are directly related to transfer functions that can be derived from the identified open-loop helicopter dynamics. For example, the closed-loop transfer function describing the roll angle response to the reference roll angle command is written

$$T_{\phi,\phi_{ref}} = \frac{G_{\phi,\delta_{lat}} K_\phi}{1 + G_{\phi,\delta_{lat}} K_\phi} \tag{6.1}$$

where K_ϕ is the roll angle feedback gain shown in Fig. 6.1, and $G_{\phi,\delta_{lat}}$ is the transfer function describing the open-loop roll angle response ϕ to the lateral cyclic input δ_{lat}, which is derived from the state-space system shown in Eq. 3.113:

$$G_{\phi,\delta_{lat}} = \boldsymbol{C}_\phi \left[s\boldsymbol{I} - \boldsymbol{A}\right]^{-1} \boldsymbol{B}_{lat} \tag{6.2}$$

where $\boldsymbol{I}$ is the unity matrix, C_ϕ is the special output matrix with unity element at the roll angle index and zero elements everywhere else (see state vector shown in Eq. 3.111), and B_{lat} is the special input matrix with a unity element at the lateral cyclic input index and zero elements everywhere else (see input vector shown in Eq. 3.112). The other transfer functions are derived by selecting the appropriate input and output matrices.

Figure 6.3 shows the frequency responses that were estimated from the flight-data superposed with those predicted by the respective closed-loop transfer functions. The agreement is good both for the magnitude and phase of the on-axis responses. In the frequencies of the coupled R-F mode, the identified magnitude does not reach the level predicted by the mode. This mismatch is related to the lack of data points in this region. The pronounced phase roll-off, which is in good agreement with the model, clearly confirms the presence of the lightly damped R-F mode. The off-axis agreement is not as good as the on-axis agreement. Most of the off-axis effects occur in the mid- to high-frequencies (5-11 rad/sec) and are related to the coupled R-F mode. These effects are well predicted by the model. The mismatch is largest in the low- to mid- frequencies (0.3-5 rad/sec). In this particular frequency range, the coherence that was computed from the flight data was well below the 0.6 threshold, indicating unreliable data.

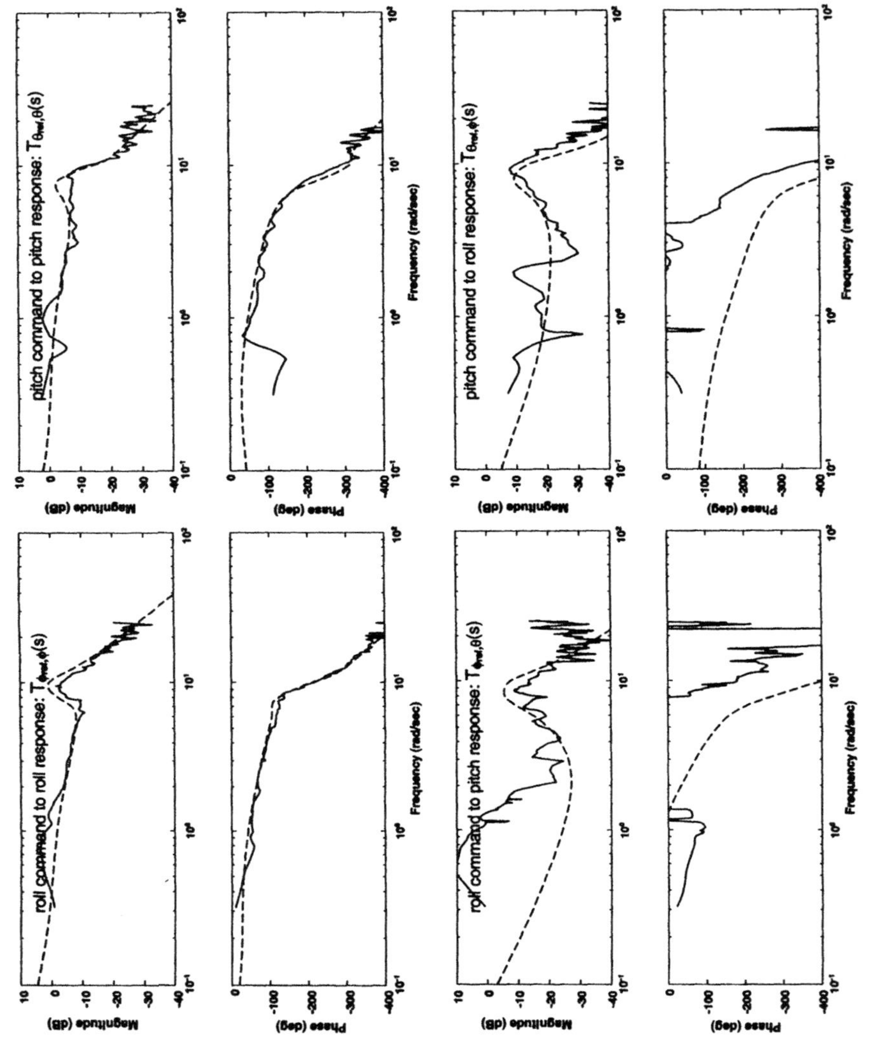

Figure 6.3. **Comparison between the main transfer functions derived from the flight experiments and the ones predicted by the model for the baseline attitude control system.**

Determination of the Loop Gain Functions

For the analysis of the closed-loop system we need to extract the loop-gain functions $L_{\phi_{ref},\phi} = G_{\phi,\delta_{lat}}K_\phi$ and $L_{\theta_{ref},\theta} = G_{\theta,\delta_{lon}}K_\theta$ from the closed-loop frequency responses. Focusing on the roll axis, the frequency responses describing the on-axis response $H_{\phi_{ref},\phi}$ can be written in terms of its respective loop gain function as follows:

$$H_{\phi_{ref},\phi} = \frac{L_{\phi_{ref},\phi}}{1 + L_{\phi_{ref},\phi}}; \tag{6.3}$$

solving for the loop gain functions, we obtain:

$$L_{\phi_{ref},\phi} = \frac{H_{\phi_{ref},\phi}}{1 - H_{\phi_{ref},\phi}}. \tag{6.4}$$

The arithmetic operation on the frequency response was performed using the utilities available in the CIFER package. The same calculation is made for the pitch axis.

Figure 6.4 shows the comparison between the model-predicted loop gain functions and the ones derived from the flight data. The good agreement between the predicted and identified loop gain function is a clear validation of the identified model and its suitability for control design.

Time Domain Verification

Besides the frequency sweep experiments, one piloted task experiment was conducted using the attitude command system. In this experiment, the pilot had to move the helicopter laterally between two traffic cones positioned four rotor diameters apart. He had to perform this translation motion as quickly as possible and bring the helicopter to a full stop within one rotor radius of the cone. This task was designed to provide some preliminary data for the development of flying qualities requirements. The data recorded during this task is used here to verify the model of the closed loop in the time domain.

Figure 6.5 shows the comparison between the flown attitude angles reference inputs and the model-predicted attitude responses. The overall agreement between actual and predicted data is good. The longer-term agreement for the roll and pitch angles is weaker. This effect is consistent with the lack of accuracy of the identified model in the low frequencies (below 1 rad/sec). This is usually not an issue for control design, because the crossover frequency of the attitude control loops is usually well above that frequency.

Identification of the Total Effective Time Delay

Time delays in the control path result in a negative phase shift and thus are an important factor for the stability of the closed-loop system. The implementation of the closed-loop system on the vehicle on-board computer involves various sources of time delays that contribute to the closed-loop dynamics. The principal sources of delays are the decimation (sampling of the sensors) and de-

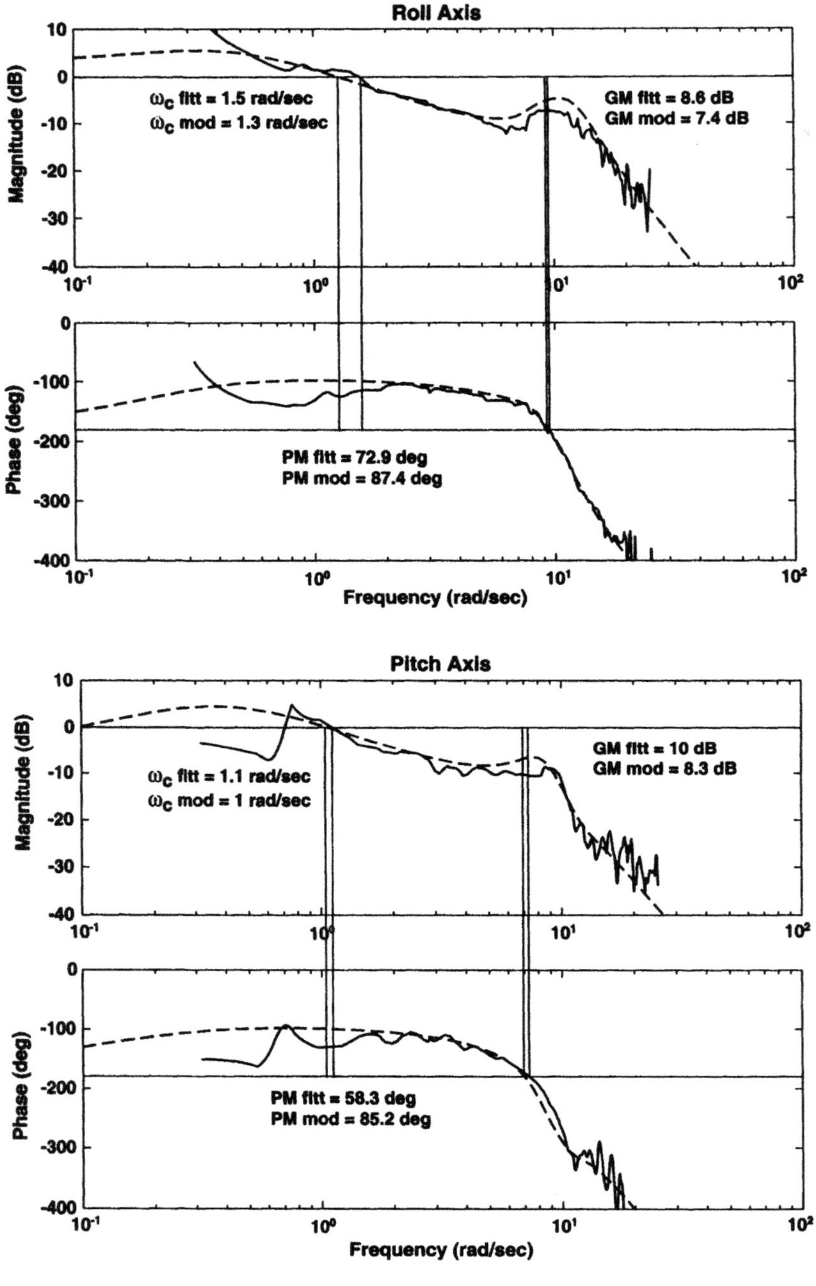

Figure 6.4. Bode plot of the loop gain functions derived from the closed-loop identification with the baseline controller, and the ones predicted using the identified model for the roll (top) and pitch (bottom) axes (including time delay). Also shown are the phase margin and the gain margin achieved with this control system. Notice that the resonance peak is located precisely at the ω_{180} frequency and thus limits the gain margin.

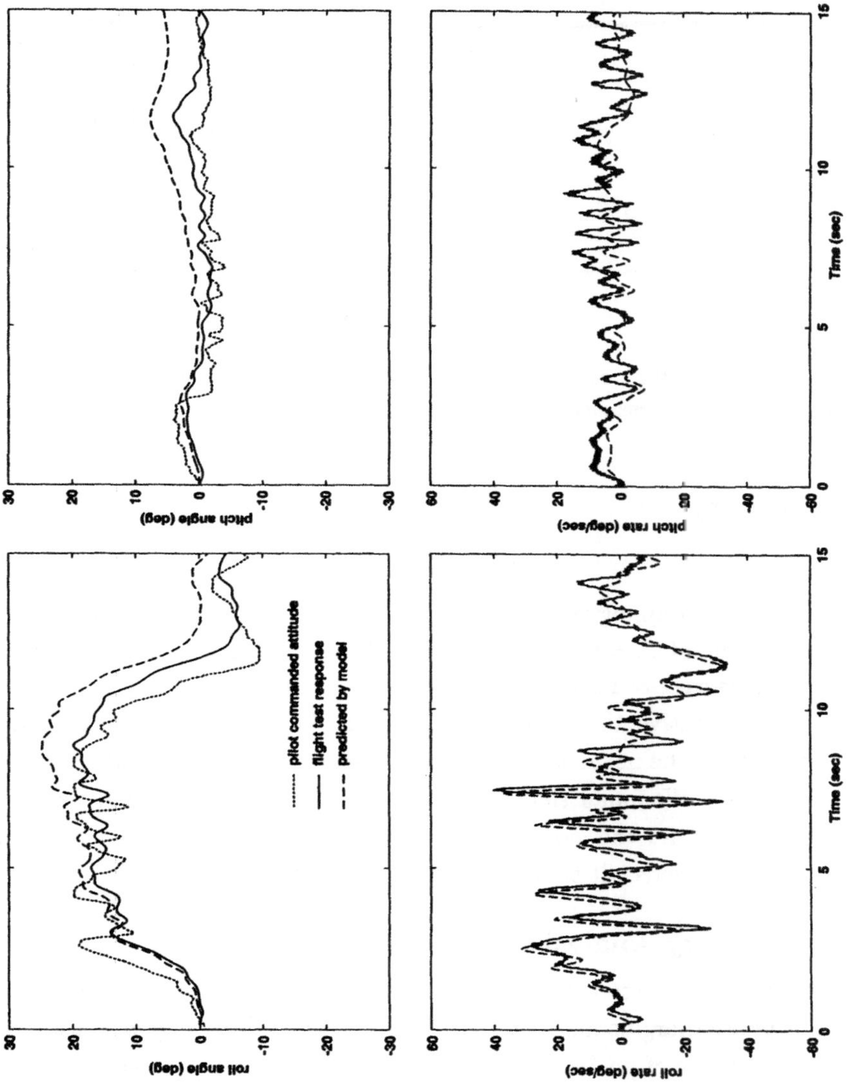

Figure 6.5. Time domain comparison between the flight-test data and the responses predicted by the model for the baseline controller. The pilot attitude command is indicated by a dotted line, the flight data by a solid line, and the model-predicted response by a dashed line.

lays related to the digital computation. Data sampling is usually modeled as a zero-order hold (ZOH) [84], which produces an equivalent delay that can be

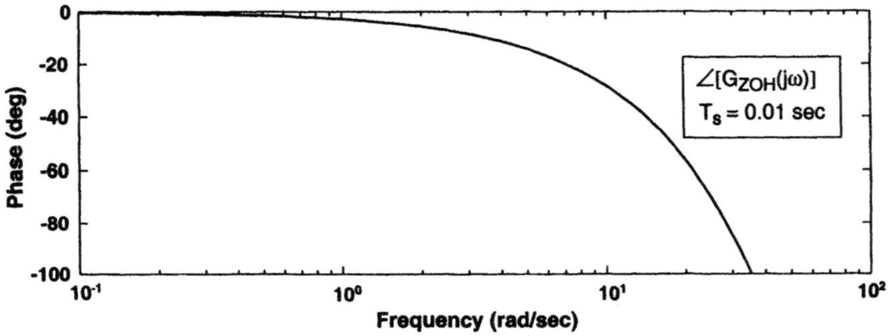

Figure 6.6. Effect of the zero order hold (ZOH) on the system's phase response, where the effect of the ZOH is approximated by a pure delay of half the system's sampling time T_s.

approximated by:

$$\angle[G_{\text{ZOH}}] = \frac{T_s}{2}s, \tag{6.5}$$

where T_s is the sampling time. Figure 6.6 depicts the ZOH phase response for a sampling time of $T_s = 0.01$ sec (100 Hz). Note that such a delay contributes to about 30 degrees of negative phase shift at 10 rad/sec.

The total effective time delay occurring in the system can be accurately determined from the mismatch between the estimated and the model-predicted phase response. We found that a total delay of twice the sampling time (0.02 sec) was required for the model to best match the flight data. Notice the near perfect phase response fit shown in Fig. 6.4.

2.2 Stability Analysis

To determine the stability margins of the baseline control system, we used gain and phase margins analysis [76]. The crossover frequency was also evaluated as a measure of performance (more metrics will be presented later in this chapter). Figure 6.4 shows the Bode plots for the lateral and longitudinal loop gain functions with the phase and gain margins (denoted by PM and GM) and the crossover frequencies ω_c indicated.

Table 6.1 shows the crossover frequencies and phase and gain margins derived from the model-predicted and identified loop gain functions. The agreement between the two is generally very good. A difference can be found for the pitch phase margin of the baseline controller: 58.3 degrees vs. 85.2 degrees for the predicted value. Looking at the Bode plot (Figure 6.4), we see that at the crossover frequency (1.1 rad/sec), the phase presents a local drop that belongs

Table 6.1. **Comparison between the actual and predicted crossover frequencies and stability margins for both the baseline PD controller (manually tuned) and the compensated PD controller (same feedback gains as baseline but with notch filter).**

		ω_c (rad/sec)		PM (deg)		GM (dB)	
		model	flight	model	flight	model	flight
baseline	roll	1.3	1.5	87.4	72.9	7.4	8.6
	pitch	1	1.1	85.2	58.3	8.3	10
compensated	roll	1.2	1.3	78.1	76.5	18.7	18
	pitch	1	1	73.9	73	19.9	13

to a series of phase ripples that are believed to be caused by the insufficient amount of data.

Based on the general specifications for flight control design (MIL-F-9490D [1]), which require a phase margin of 45 degrees and a gain margin of 6 dB, the baseline system lacks adequate phase margin. The phase margin could be increased by raising the bandwidth; however, as can be observed from the loop gain function, this would be detrimental for the gain margin. This is a typical situation where the gain and phase margins determine the maximum bandwidth. The bandwidth limitation is particularly pronounced for our RUAV because the critical frequency ($\omega_{180,lat} = 11.4$ rad/sec and $\omega_{180,lon} = 8.4$ rad/sec) coincides with the natural frequency of the lightly damped R-F mode. The peak in magnitude of about 14 dB ($-\log_{10}(\zeta = 0.2)$), due to this lightly damped mode, translates directly to a reduction in the gain margin.

This analysis shows that, with the current controller architecture, no further performance increase can be attained, i.e. the PD control system already exploits all the available performance. Typical attitude control systems [56], in addition to the attitude feedback, also use attitude rate feedback. However, because of the high-frequency gain and the additional phase lag caused by a differential element, an attitude rate feedback would not address this limitation.

From this analysis, we conclude that the light damping in the coupled R-F mode due to the stabilizer bar constitutes a performance and robustness limitation for classical feedback control systems.

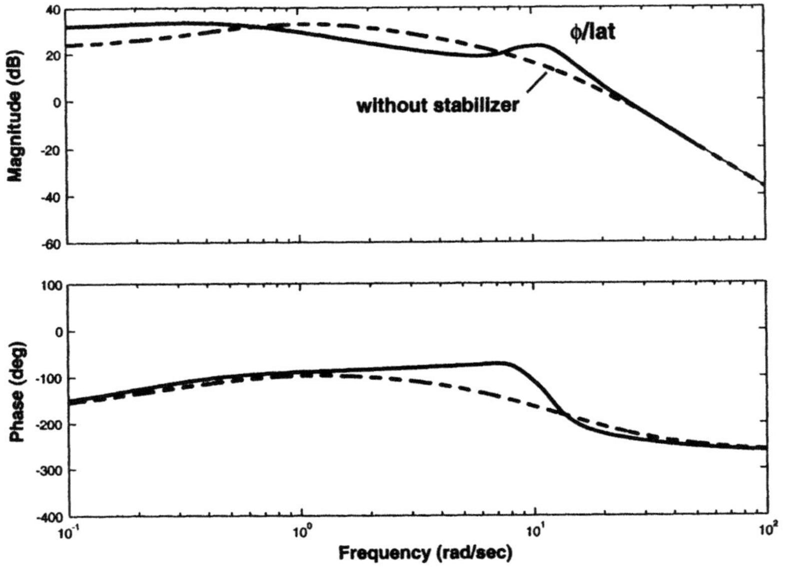

Figure 6.7. Comparison between the natural roll response (solid) and the roll response without the stabilizer (dashed). Note how the resonance peak disappears without the stabilizer.

2.3 Compensation of Attitude Dynamics

An accurate model opens a wide range of possibilities for design and optimization of control systems, and several design methods could potentially address such a limitation. In the following, we introduce a solution to increasing the performance of the PD control system, which could not otherwise produce a higher performance.

One way to address the performance limitation would be to increase the damping in the R-F mode, or compensate for the lightly damped mode. Without a stabilizer bar, the R-F mode is usually critically damped. This is visible in Figure 6.7, which shows the ϕ/δ_{lat} frequency response generated from the identified model for cases with and without a stabilizer bar. Hence, the limitation could be addressed by removing the stabilizer bar. However, as we saw in Chapter 5 the stabilizer bar is required for piloted flight, and many RUAV projects still rely on manual capabilities for backup. In the following, we explore a less radical solution.

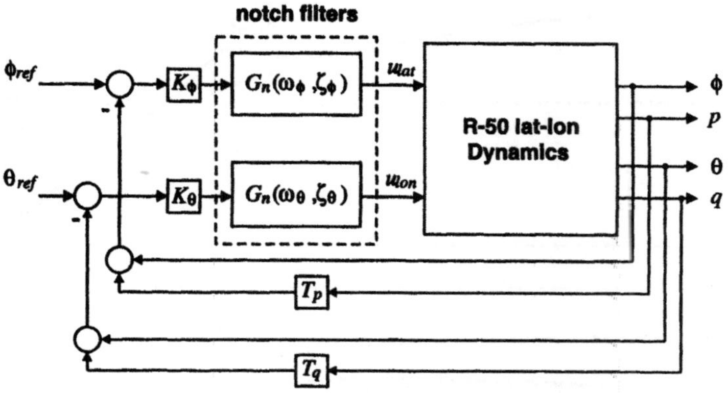

Figure 6.8. Blockdiagram of the PD attitude control system including the notch filters

Notch Filter Compensation

The lightly damped mode in the attitude dynamics is fully characterized by its natural frequency and damping ratio. We have already seen that these characteristics are defined by fundamental parameters of the rotorcraft. Namely, the natural frequency is determined by the flapping spring derivatives (Eq. 5.24; L_b for the roll axis and M_a for the pitch axis), and the damping ratio is determined by the flapping spring derivatives with the equivalent rotor time constant (see Eq. 5.25).

Therefore, a second-order notch filter of the form

$$G_{\text{notch}} = \frac{s^2 + 2\zeta\omega s + \omega^2}{s^2 + 2\omega s + \omega^2}, \tag{6.6}$$

tuned with the precise system modal characteristics (natural frequency and damping), should adequately compensate for the resonance peak and result in better behaved attitude dynamics. The filter is placed in the control path of the attitude loops as shown in Fig. 6.8.

For example, for the roll dynamics, the natural frequency of the R-F mode is $\omega_{roll} = 11.88$ rad/sec, and its damping ratio is $\zeta_{roll} = 0.22$. Figure 6.9 shows the effect of the second-order notch filter on the roll attitude dynamics. As can be seen, the resonance peak due to the stabilizer bar is well attenuated by the notch filter.

The notch filter should allow for significantly higher attitude gains while providing sufficient gain and phase margin, thereby allowing for a higher bandwidth without exacerbating the coupled R-F mode. Note that this solution

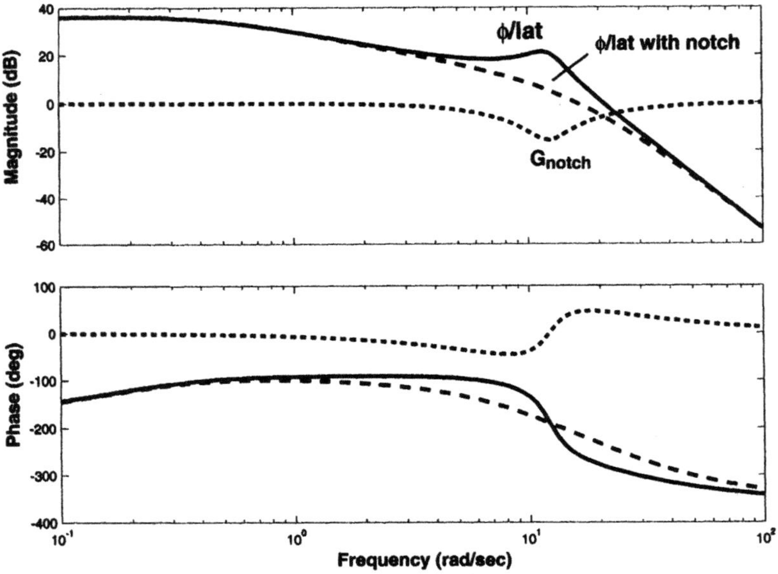

Figure 6.9. Comparison between the original roll response (solid) and the roll response with a notch filter (dashed). Also shown is the frequency response of the notch filter (dotted). Note the compensation of the resonance peak by the notch filter.

does not cancel out the intended effects of the stabilizer bar (pseudo attitude rate feedback effect and the increase in rotor damping), but compensates only for the effect of the stabilizer on the coupled R-F mode. Note that the notch filter also improves the handling qualities for piloted flight.

Flight-Test Evaluation of the Notch Filter

To verify the effectiveness of the notch filter in attenuating the resonance peak of the R-F mode, we proceeded with a closed-loop identification of the baseline PD controller fitted with the notch filters as shown in Fig. 6.8 (note that in the baseline controller, the rate response feedback gains T_p and T_q are zero). The procedure is identical to the one used for the uncompensated system. From the estimated loop gain we determined whether the notch filter produced the predicted increase in gain and phase margins.

Figure 6.10 shows the comparison between the estimated and model-predicted loop gain functions. Overall, the agreement of the loop gain functions is good, and the agreement between the actual and model-predicted crossover frequencies and stability margins is also good. Only one discrepancy can be seen: the

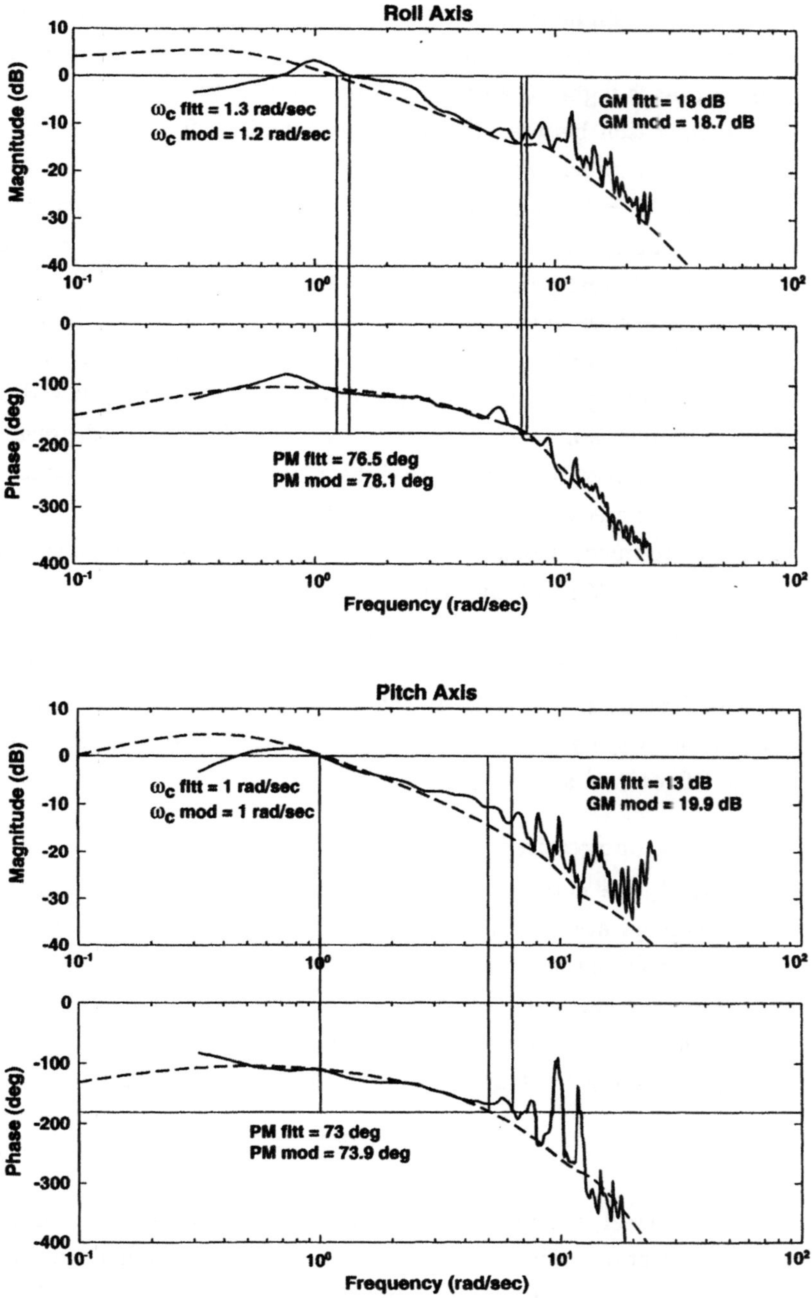

Figure 6.10. Bode plot of the loop gain functions derived from the closed-loop identification with the compensated PD controller, and the ones predicted using the identified model for the roll (top) and pitch (bottom) axes. Also shown are the phase margin and the gain margin achieved with this control system. Notice the absence of the typical resonance peak and the dramatic increase in stability margins. (see Fig. 6.4 for comparison).

model predicts a roll gain margin of 19.9 dB, while the flight data shows a 13 dB value. The difference is the result of a pitch response magnitude obtained from the flight data that is higher than the one predicted by the model.

From the comparison between the loop gain functions for the baseline and the compensated controller, we observe that with a notch filter tuned to the characteristics of the R-F mode, a dramatic increase in gain margins can be achieved. For example, for the roll axis, the gain margin goes from 8.6 dB to 18 dB. This increase in margin can be used toward increasing the feedback gains and improving the performance. Based on the slope of the loop gain function in the crossover region (-10 dB/decade), an increase in the feedback gain of half the improved gain margin would push the crossover frequency half a decade higher. One downside is a decrease in phase margin of about 10 degrees. An increase in gain margin of 10 dB far outweighs a 10-degree loss in phase margin (this tradeoff can be assessed by looking at a Nichols plot).

The piloted experiments using the task with the notch filter described earlier were also conclusive. Without the notch filter, the pilot can easily excite the coupled R-F mode, which results in a very oscillatory attitude. With the notch filter, this tendency disappears entirely. Note that the notch filter does not prevent the R-F mode from being excited by wind gusts. A high-bandwidth notched PD control system will be better at rejecting disturbances than the slow baseline PD control system.

Robustness to Changes in System Characteristics

To be practical, the notch filter must be sufficiently robust to changes in the helicopter configuration or operational conditions. In the following, we will determine how sensitive the key parameters involved in the coupled R-F mode are to changes in system characteristics.

Recall that the natural frequency of the R-F mode, for example ω_{roll} for the roll axis, is a function of the lateral flapping spring constant L_b

$$\omega_{\text{roll-flap}} \approx \sqrt{L_b}. \tag{6.7}$$

The value L_b is inversely proportional to the roll moment of inertia (see Eq. 3.77), which is a configuration variable subject to variations (e.g. changes in the payload). However, a 20% change of the helicopter moment of inertia – which is a very large variation – produces only a 10% change in the natural frequency of the R-F mode. As can be seen in Figure 6.11, shifting the notch filter frequency by 10% of the nominal frequency only marginally changes the attenuation of

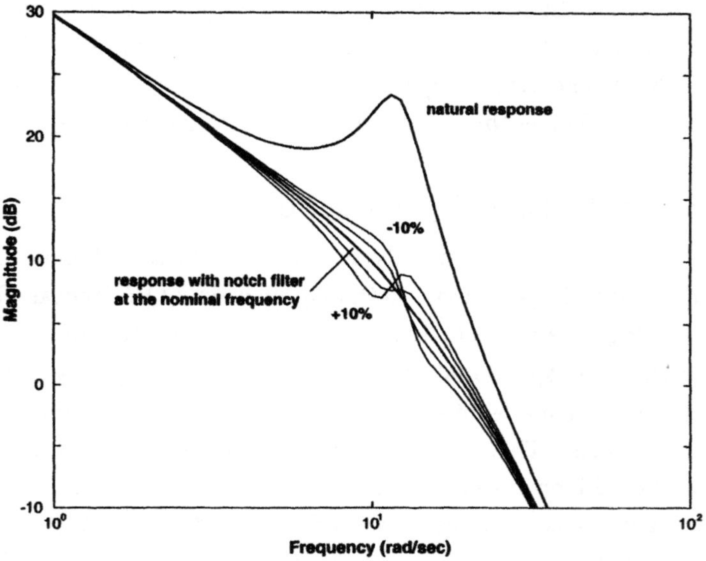

Figure 6.11. **Effect of variation in the natural frequency of the coupled R-F mode (±10 %) on the attenuation of the resonance peak achieved by the notch filter.**

the resonance peak. The flapping spring is also a function of the rotor thrust. For a less agile helicopter such as the R-50, the thrust does not change dramatically during flight. For an acrobatic vehicle such as the X-Cell, the thrust can cover a large range of values (thrust between -1 to 2 times the vehicle weight). Acrobatic rotorcraft usually have stiff rotor heads; hence, their flapping spring derivative is dominated by stiffness. For example, for a 2-g turn with the X-Cell helicopter, Eqs. 3.77 and 6.7 predict that the natural frequency will change by about 12 %.

The damping ratio of the coupled R-F mode ζ_{roll} is a function of the normalized rotor spring coefficient L_b, and the rotor time constant τ_f (effective time constant for a rotor with stabilizer bar)

$$\zeta_{\text{roll-flap}} \approx 1/(2\tau_f\sqrt{L_b}). \tag{6.8}$$

The rotor time constant is determined largely by the rotor speed Ω and the blade Lock number γ, as shown by Eq. 4.19. The rotor speed is usually kept constant by an engine governor. The blade Lock number can be influenced by the rotor inflow (see effective Lock number Eq. 4.22). This effect, however, is small. Hence, the variations in operating conditions are not expected to have large effects on the rotor time constant. From our identified cruise model, we observe less than 10 percent change in the parameters of the R-F modes between the hover and cruise flight condition. Therefore, a single notch filter design should be effective throughout most of the flight envelope.

3. Control System Optimization

A model can be used in essentially two ways for control design. In model-based design methods (synthesis methods) such as LQR (Linear Quadratic Regulator) or H_∞, the feedback law is computed from algebraic equations involving the the state-space matrices representing the plan's dynamics. The approach involves the model of the plant indirectly to optimize the parameters of the control system. This is the case for CONDUIT, which uses a model to simulate the closed-loop system and evaluate its performance via various forms of metrics against specifications. The difference between actual and desired performance drives the optimization of the feedback parameters. In the following we will use this optimization tool with our identified model to illustrate the performance that can be attained with the notch filter enhanced PD controller.

3.1 CONDUIT Optimization Framework

CONDUIT is an interactive control design framework developed for flight control optimization. Flight control optimization is exceptional in that a wide range of performance specifications that pertain to criteria ranging from simple performance objectives to more complex piloted handling qualities may be accounted for in a systematic way. The integration of all the specifications in a single optimization process saves time, allows a comprehensive understanding of the design process, and enables the management of design tradeoffs.

CONDUIT combines elements from Matlab [55] with a multi-objective optimization engine. A graphical user interface incorporates in Matlab's Simulink and series of custom tools for the representation of the specification and the monitoring of the optimization process. A description of this system is available in [22]. The model of the vehicle dynamics and the detailed control system architecture, including switching logic for different control modes, are represented in Simulink. The controller architecture is generally prescribed by the user, who also determines which individual parameters are used for the optimization.

The specifications used by CONDUIT are divided into constraints and performance objectives. Constraints are used for stability specifications (e.g. eigenvalues of the closed-loop system in the left-half-plane) and usual saturations occuring in the plant (e.g. actuators). The performance objectives are used to drive the design towards a target performance. Such objectives can take a range of forms.

For evaluation, most of these objectives follow a three-level scale following the Cooper and Harper handling qualities scheme. Level 1 corresponds to the target value range; Level 2 corresponds to a performance with minor to significant deficiencies; Level 3 corresponds to performance beyond major deficiencies. Note that in this context, these levels are not necessarily related to handling qualities. Basic performance objectives include bandwidth, crossover frequency, and damping ratio. More complex objectives are based on handling qualities criteria such as those found in the Handling Qualities Requirements for Military Rotorcraft ADS-33D [2]. Some of these objectives are available in a CONDUIT specification library. The user can also design his or her own specifications using the functions available in Matlab.

The evaluation of the design is based on the performance predicted by the model of the vehicle dynamics. At each iteration, all the specifications are evaluated and the results converted into a cost which is used to drive the optimization.

3.2 Attitude Control Optimization

For the optimization we use the notch filter-enhanced PD attitude control architecture shown in Figure 6.8. Note the added rate feedback loop. All the feedback gains were selected as optimization parameters. To provide additional degrees of freedom, we also included the damping ratio of the notch filter. The optimization parameters are:

- K_ϕ, K_θ: lateral and longitudinal attitude gains
- T_p, T_q: ratio of rate to angle feedback
- ζ_ϕ, ζ_θ: notch filter damping ratio

Attitude Performance Specifications

In hover flight, the role of the attitude control system is mainly stabilization, tracking small position changes, and disturbance rejection, i.e. holding a position in gusty conditions. We will not try to attain a target performance. In qualitative terms the goal is to optimize the system for three levels of performance measured in terms of the speed of response (rise time), and we want the responses to be overshoot and oscillations free. The performance should range from what is attained with the baseline system, to the minimum speed of response which can be achieved without violating the stability margins presented next.

To measure the performance attained during the optimization we use basic control metrics. More elaborate metrics used in the attitude flying qualities that could be used in such a design task will be presented in the next section. Figure 6.12 shows the specifications used in CONDUIT for the optimization. They consist of

- *Eigenvalue location*: constrains the real parts of all closed-loop system's eigenvalues to be less than or equal to zero, thus ensuring that all the dynamics are stable or neutrally stable.

- *Gain and phase margins*: These quantities determine the robustness to changes in the system. For piloted operations, Level 1 requires a minimum of 45 degrees of phase margin and 6 dB (taken from MIL-F-9490D). Since we are not considering handling qualities, these margins will ensure good robustness to unmodeled dynamics and changes in the dynamics near hover flight.

- *Bandwidth*: Attitude control bandwidth (-3 dB frequency) determines the speed of response of the vehicle attitude to an attitude reference input. For piloted control, a high bandwidth is necessary for good handling qualities. For comparison, for full-scale rotorcraft, the specification from ADS-33D for a Level 1 handling quality for the tracking task is 2 rad/sec.

- *Crossover frequency*: limits the crossover frequency, which is used here as a measure of the control activity. The Level 1 upper limit is set at 10 rad/sec. Actuator saturation was used later to determine the maximum attitude and angular rates that can be attained with the different controllers (see Table 6.8).

- *Frequency response envelope*: This specification was designed to precisely shape the frequency response of the closed-loop attitude response. The envelope penalizes a resonance peak in the roll and pitch attitude responses and thus limits response overshoot and oscillations caused by the coupled R-F mode. At the same time, the horizontal position of the envelope is used together with the bandwidth specification to adjust the control system bandwidth. The CONDUIT bandwidth specification pushes the bandwidth from below, and the frequency response envelope sets an upper limit.

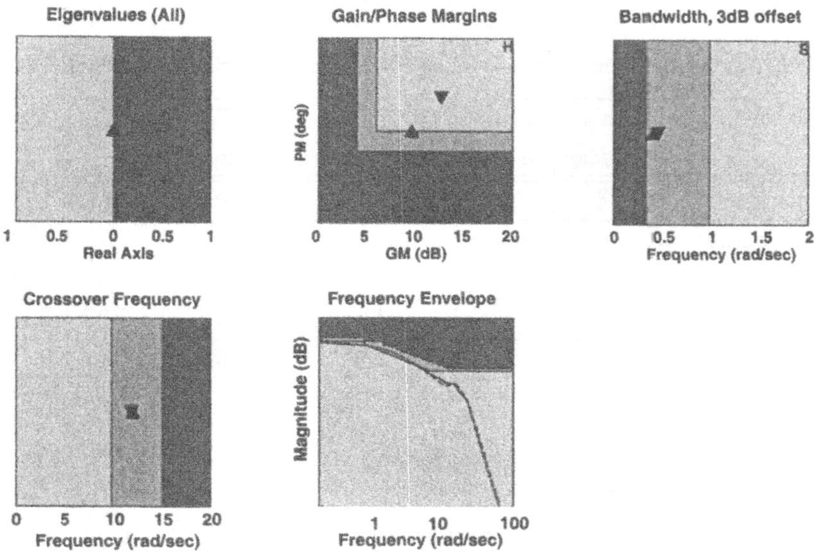

Figure 6.12. **Specification window from CONDUIT showing the specifications used for the attitude control optimization. The light gray area represents the Level 1 target area. The results are shown by the triangle icons (upright: pitch; inverted: roll). The frequency response envelope shows the frequency response boundaries and the resulting roll (dashed line) and pitch (solid line) frequency responses.**

In this example no specific care was taken to minimize cross-coupling effects or disturbance rejection. The R-50's roll-pitch coupling is most severe at the frequencies of the R-F mode (5-11 rad/sec; See Figure 4.4); limiting the excitations of that mode with the notch filter will limit this source of coupling. High low-frequency gain will address the low-frequency coupling. In CONDUIT cross-coupling and disturbance rejection could both be addressed using time domain or even frequency domain specifications. The time domain specification will allow to prescribe a limit in the magnitude of the attitude off-axis response. The frequency response specification would prescribe the peak magnitude in the cross-axes transfer functions. The same approach could be used for the disturbance rejection. For very precise specification of the response characteristics up to higher bandwidth, high-order frequency domain design methods like H_∞ can be advantageous [92].

Table 6.2. Controller Gains for the baseline system and the optimized systems.

	K_ϕ % /deg	K_θ % /deg	T_p % / (deg/sec)	T_q % /(deg/sec)	ζ_ϕ	ζ_θ
baseline	4	4	0	0	1	1
compensated	4	4	0	0	0.22	0.2
slow	8	17	42	60	0.22	0.2
medium	13	20	32	40	0.07	0.14
fast	17	21	20	33	0.11	0.13

Optimization Results

We optimized the control system for three levels of performance by moving the frequency response envelope along the frequency axis, thus changing the upper bandwidth limit. Table 6.2 shows the controller gains for the baseline system (manual tuning with only angular feedback and with no notch filter), the gain for the compensated system (same as baseline but with notch filter tuned to match the natural characteristics of the coupled R-F mode), and the gains obtained during the performance optimization.

Figure 6.13 shows the simulated roll time responses for a roll reference step of 1 degree. Large amplitude changes will be considered in the next section. First, we see that all notch-enhanced controllers do not exhibit the short period oscillation from the R-F mode exhibited by the baseline PD controller. The responses of the optimized controllers show an increase in speed proportional to the bandwidth requirement. The bandwidth goes from $\omega_{-3dB} \cong 1.3$ rad/sec for the baseline system to $\omega_{-3dB} \cong 6.11$ rad/sec for the fastest controller. The rise time for the medium and fast controllers are 0.96 sec and 0.64 sec, respectively, compared to 1.22 sec for the baseline (see Table 6.3). As can be seen in Table 6.2, the increase in performance is achieved through a fivefold increase in the roll angle feedback gain ($K_\phi = 21\%$/deg for the fastest controller compared with $K_\phi = 4\%$/deg for the baseline system).

3.3 Velocity and Position Performance Margins

The performance of the attitude controller prescribes the performance that can be attained by a higher level velocity or position controller. This can be illustrated through the Bode plots of the loop gain functions calculated for the

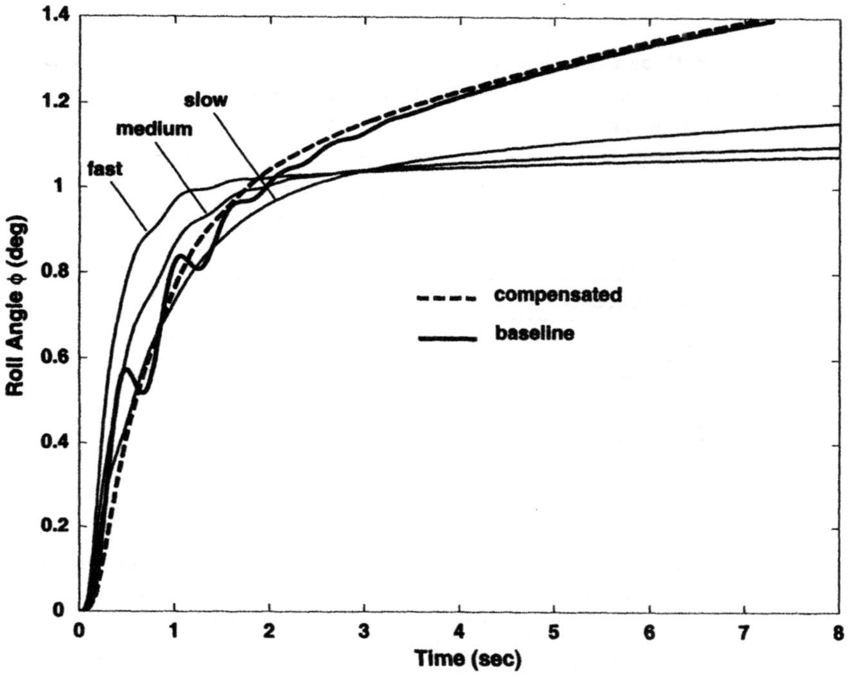

Figure 6.13. Comparison between the simulated responses to a roll angle reference step of the different attitude control designs. Shown are the results, for the baseline system, the compensated system, and the three optimized systems (slow, medium, fast; see Table 6.2). Notice the short-term oscillation in the response of the baseline system and how it is successfully addressed by the notch filter.

velocity $L_{v,\phi_{ref}}$ and position $L_{y,\phi_{ref}}$ with the different attitude control loops closed.

The Bode plot of the loop gain function for the lateral velocity (Fig. 6.14), shows that a faster attitude controller provides more phase and gain margin (PM= 168 deg for the fast design vs. PM= 125 deg for the baseline system). The increased margin allows for a higher-velocity control bandwidth, while maintaining sufficient stability margins to preserve robustness.

The same conclusion can be reached for the position control (Fig. 6.15). For the position control, ω_{180} goes from 0.37 rad/sec for the baseline system to 0.67 rad/sec for the fastest design. The gain margin increases from zero to almost GM= 10 dB for the fastest design. The phase margin for the fastest attitude control design is about 15 degrees higher than that of the baseline system.

Table 6.3. Bandwidth and rise time for the different controllers. The Level 1 bandwidth requirement for target acquisition and tracking tasks is 2.5 rad/sec and 2 rad/sec for, respectively, the roll and pitch axes.

	ω_{-3dB} rad/sec	t_r sec	ω_{gain} rad/sec	ω_{phase} rad/sec
baseline	1.3	1.22	0.13	10.39
compensated	1.5	1.05	5.4	7.82
slow	1.3	1.43	5.69	12.05
medium	2.2	0.96	9.95	10.83
fast	6.1	0.64	10.93	10.44

Table 6.4. Crossover frequency and gain and phase margins for the different controllers

	roll			pitch		
	ω_c rad/sec	GM dB	PM degrees	ω_c rad/sec	GM dB	PM degrees
baseline	1.3	7.4	87.4	1.1	8.3	85.2
compensated	1.2	18.7	78.1	1	19.8	73.9
slow	9.3	13.1	62.6	7.3	13.1	52.6
medium	9.6	13	60.1	7.8	11.9	59.9
fast	11.9	12.6	61.6	12	9.6	45.2

These results are meaningful independent of the method by which the velocity or position loops are closed, i.e. they are equally valid for a human pilot and a classical or a modern controller. However, one method will be better suited than another for exploiting the performance under a particular attitude controller.

4. Criteria for Specification and Evaluation of Performance

Typical control specifications like the ones we used for the optimization of the attitude control system are generally not sufficient to guarantee the effective execution of a task or even a particular maneuver with the vehicle. Specifica-

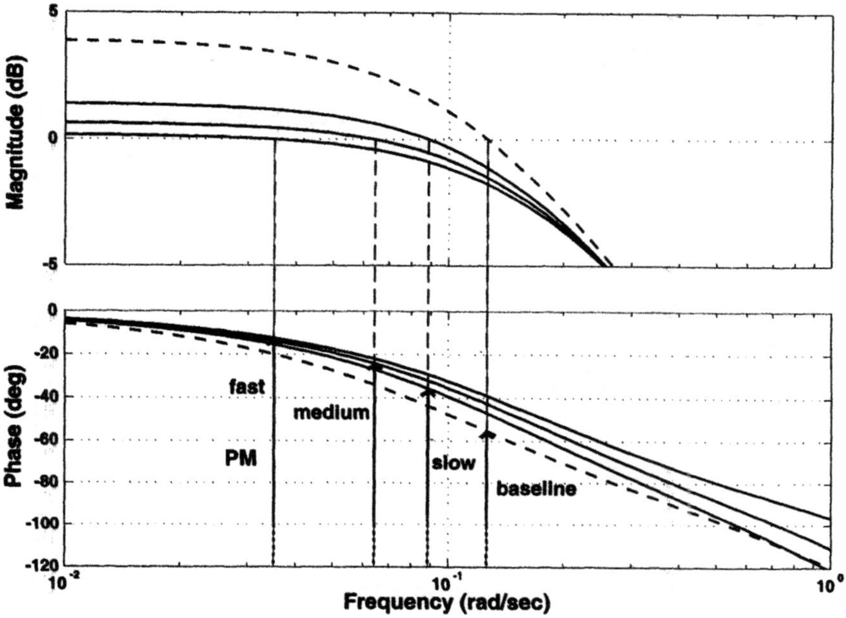

Figure 6.14. Comparison of the velocity loop gain function for the different attitude control systems. Shown are the baseline system (baseline - dashed) and the three optimized systems (slow, medium, fast) with notch filter in order of increasing bandwidth. Notice how the phase margin increases with faster attitude control.

tions used to design and evaluate a flight control system need to encompass aspects related to the task, the flight conditions, the control modality (how the operator controls the various vehicle axes, i.e. whether the system is remotely operated in manual "piloted" mode or through a stability augmentation such as an attitude hold system), and the interface (in remote operation how the information about the vehicle state is presented to the operator).

In the following, we will look at the notion of flying qualities (FQ) in manned vehicles and discuss their extension to unmanned vehicles. Then we will present the key attitude flying qualities metrics used for attitude hold systems. To illustrate their usage we will apply them to our attitude controllers. This will also give us a sense of why and how such criteria could be extended to miniature unmanned rotorcraft. First, we give an illustrative example.

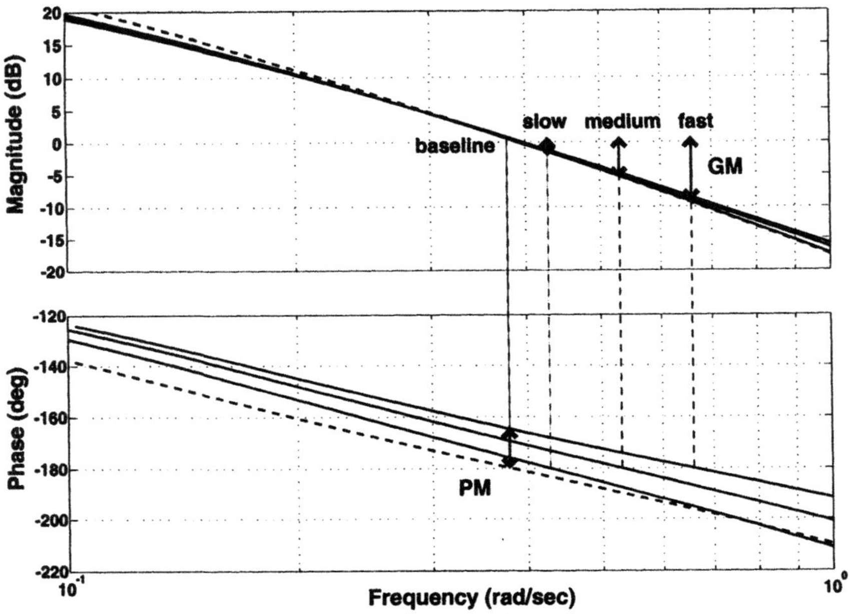

Figure 6.15. Comparison of the position loop gain functions for different attitude control systems. Shown are the baseline system (baseline - dashed) and the three optimized systems (slow, medium, fast) with notch filter in order of increasing bandwidth. Notice how the phase and gain margin increases with faster attitude control.

4.1 The Notion of Flying Qualities

The experiment described earlier for the time domain verification is a good example of the necessity of flying qualities requirements. The pilot was controlling the helicopter via the remote control unit while standing in immediate proximity of the vehicle. With the baseline attitude PD controller, it took him about 8 seconds to travel the distance and end in precision hover within one rotor radius of the target. Compared to the approximately 4 seconds the maneuver required without augmentation, it was a poor performance that nevertheless required a high level of attention from the pilot. The pilot's main complaint was that he felt that the attitude control system was resisting his commands –a result mainly of a poor bandwidth. This led him to overcontrol the helicopter and resulted in the short-term oscillations from the R-F mode that are visible in Fig. 6.5.

Manned Vehicles

The notion of flying qualities for manned aircraft was introduced by Cooper and Harper as "those qualities or characteristics of an aircraft that govern the ease and precision with which a pilot is able to perform the tasks required in support of an aircraft role" [15]. For an objective evaluation of an aircraft's flying qualities level by a pilot, criteria are defined precisely for different tasks and conditions.

The criteria and requirements that were developed by the U.S. Army for military rotorcraft are summarized in the Aeronautical Design Standard ADS-33D [2]. The requirements are parameterized by so-called Mission Task Elements (MTE) and Usable Cue Environments (UCE). MTE represent the type of tasks that need to be accomplished. These are given explicitly (e.g. shipboard landing) or in form of a maneuver (e.g. rapid transition to precision hover). The UCE represent the levels of cues available to the pilot to orient him- or herself with respect to attitude and translation. The requirements for a particular MTE will change in function of the UCE. For example, holding a position in hover will me more demanding in poor visibility and thus will require a better behaved vehicle.

If deficiencies in flying qualities are detected, design modifications are implemented. In the past, these involved modifications to the airframe and flight mechanics. Since the advent of active control technologies, flying qualities can be synthesized. To be effective, control optimization must use performance specifications that precisely reflect the handling qualities requirements. It was precisely the challenge of handling a large number of heterogeneous specifications that multi-objective control design optimization tools such as CONDUIT were developed.

Unmanned Vehicles

For the operation of an unmanned rotorcraft, the question of flying qualities is also relevant. The RUAVs FQ problem encompasses new issues that arise from the variety of ways a RUAV can be operated. Operation can range from piloted modes with various levels of augmentation to fully autonomous modes. For example, for the operation of RUAVs by an operator, the form of interface will define the types of cues available to the operator (similar to UCE). We can imagine the use of virtual or enhanced reality systems that will allow the pilots to best immerse themselves in their task. For fully autonomous operations, the

Table 6.5. Roll angle ranges used for the roll performance specification, with typical maneuver types and performance metrics used for each range.

Input Range	Typical Maneuver	Performance Criteria
small $\phi < 10°$	tracking e.g. hover, trim trajectory	bandwidth
medium $10° < \phi < 60°$	moderate maneuvers e.g. accel. decel., turns	quickness
large $\phi > 60°$	aggressive maneuvers e.g. tight turns, dash	achievable rate and angle

performance requirement will essentially determine the level of sophistication of the autonomous flight control system.

4.2 Attitude Flying Qualities Criteria

The attitude axis is used both for tracking and maneuvering. For example tracking could be holding a hover or following a precise trim trajectory; this also includes rejecting gust disturbances. Maneuvering could be accelerating/decelerating or banking to turn. Tracking usually requires small attitude excursions, while maneuvering requires moderate to larger excursions. To adequately encompass these different regimes, different specifications are used for different amplitude ranges. Table 6.5 shows the three roll amplitude ranges, typical maneuvers that are executed under these ranges, and the type of specification used to determine the performance of the attitude tracking system.

In the following, we will use the three specifications to evaluate the optimized controllers. To put the results into perspective, we compared the results with the Level 1 specifications from ADS-33D. We chose the requirements that are the most demanding – those used for target acquisition and tracking.

Bandwidth

The bandwidth is a measure of performance in high-gain closed-loop tracking tasks which require small-amplitude attitude changes. For control systems is usually defined as the frequency at which the output magnitude is 0.707 (-3dB) times the magnitude of the input (sometimes called servo-loop bandwidth).

It is determined from the closed-loop transfer function (i.e. complementary sensitivity function).

For piloted operations, another bandwidth definition, one that explicitly accounts for stability margins, is used, namely: the phase- ω_{phase} and gain ω_{gain} bandwidth. Both are derived from the loop gain function. The phase bandwidth is defined as the frequency at which the phase reaches -135 degrees; thereby insuring phase margin of 45 degrees when operating at this frequency (e.g. to compensate for physiological lag; based on models of human pilot behavior [57]). The gain bandwidth is defined as the frequency where the magnitude is 6dB higher than the magnitude at the ω_{-180} frequency; insuring a 6dB gain margin when operating at ω_{gain}. The smallest value between ω_{phase} and ω_{gain} is used.

Table 6.3 shows the results obtained from the evaluation of our different attitude control systems. To see the correlation between bandwidth and speed of response we included the rise time t_r. The traditional bandwidth ω_{-3dB} gives results that are consistent with the designs and rise times. The phase bandwidth ω_{phase} gives results around 10 rad/sec. It gives the highest bandwidth for the slow design. This is believed to be caused by the lightly damped R-F, which strongly influences the phase roll-off (note the natural frequency $\omega_{roll} = 11.88$ rad/sec for the roll R-F mode). The gain bandwidth obtained for the baseline controller is very small (0.13 rad/sec) compared with the slow system (5.69 rad/sec), even though the two systems get similar values for the rise time. This is due to the presence of the lightly damped R-F magnitude peak in the baseline controller, which puts the 6dB point at a much lower frequency. Otherwise, for the compensated and optimized systems, the gain bandwidth follows rise time consistently. However, the values obtained for gain bandwidth are high compared to the traditional bandwidth. The application of gain and phase margin to helicopters with a stabilizer bar should be performed carefully.

The Level 1 bandwidth requirement for target acquisition and tracking tasks is 2.5 rad/sec and 2 rad/sec for the roll and pitch axes, respectively. If we use the traditional definition, only the fastest design satisfies the requirement. However, values from the manned requirements are likely not meaningful for piloted unmanned operations. This conclusion was already made in a study performed by Peters *et al.* [70], where the authors determined the longitudinal flying qualities requirements for a lightweight fixed-wing UAV. They concluded that a small-scale aircraft requires significantly higher bandwidth to attain the same level of flying qualities as a full-scale aircraft.

The 45 degrees and 6dB values used in the phase and gain bandwidth are based on extensive studies of the behavior of the pilot in the feedback loop [57]. Such considerations will change with operations through an interface on the ground. In piloted operations of unmanned vehicles, the pilot (or operator) has fewer cues than in a manned vehicle. Sitting in the aircraft, a pilot has better position and attitude cues than when operating from outside, in particular due to the added cues from the vestibular system and the sensation of acceleration through the body. Therefore, taken into the current FQ framework, remote operations can be considered as degraded UCE and hence requiring higher FQ requirements. This tendency will be reinforced when operating through interfaces such as displays. Effects of scale will likely also influence the requirements. With smaller vehicles being naturally more agile, it should not be a problem to produce high performance. However, the increased agility will be more difficult to exploit by a human operator. The challenge will be in the development of effective operator interfaces that permit exploitation of the vehicle performance to the fullest with a minimum of operator/pilot workload. Therefore, the emphasis will be less on the vehicle and more on the operator stations.

Attitude Quickness

Quickness is a performance measure for moderate-amplitude attitude changes. It is defined as the ratio of peak roll (or pitch) rate to change in roll (or pitch) attitude.

$$\text{roll attitude quickness} = \frac{p_{pk}}{\Delta\phi}. \tag{6.9}$$

Quickness is directly related to bandwidth, but it gives a good measure of the agility of the aircraft for larger attitude changes. It can be understood as the amount of angular travel required to reach a certain angular rate. Quickness is high when the peak angular rate is high and the angular travel required to reach that peak rate is small. Pilots correlate high quickness with crispness of the attitude response to their commands.

Quickness is evaluated through simulating (or flight testing) the vehicle response to a moderate commanded attitude change. For our evaluation, we used an attitude command of 15 degrees. Figure 6.16 shows the step responses used to derive quickness, and Table 6.6 lists the results obtained for the different control systems.

The results are consistent with what is expected from the designs, i.e. a higher quickness is attained for the faster control systems, except for the baseline system, which exhibits a quickness that surpasses the faster designs. This

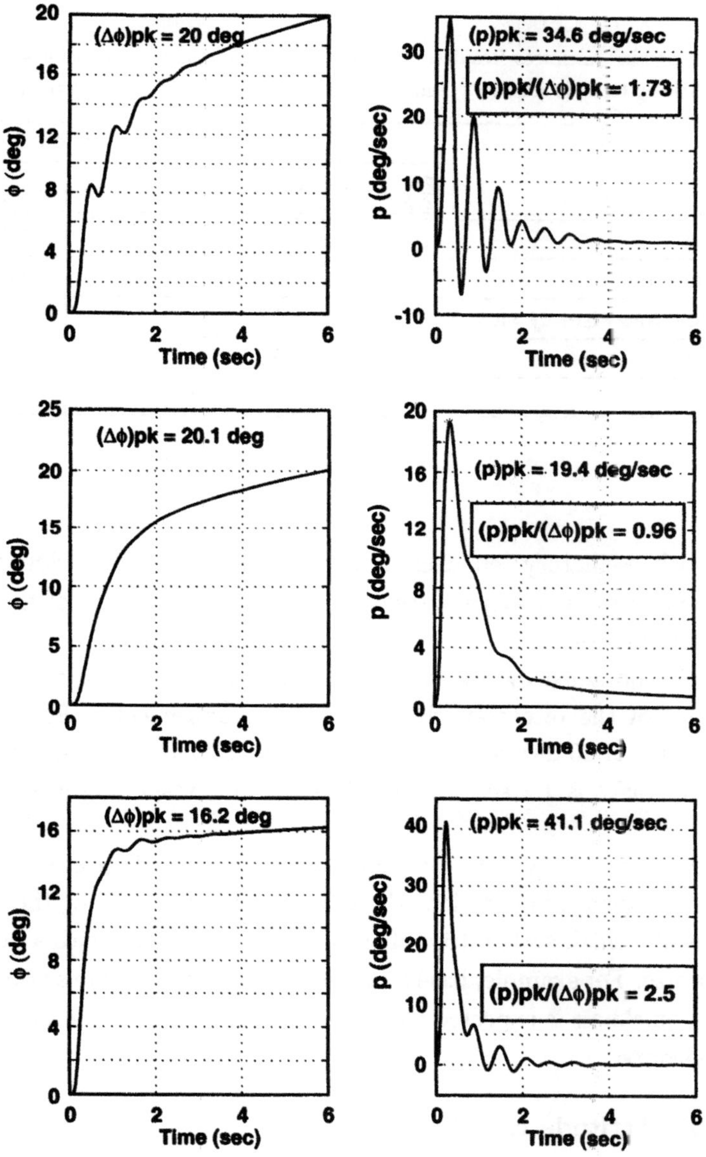

Figure 6.16. **Time histories used for the calculation of roll attitude quickness. Top row: baseline system; bottom row: fast optimized system. Notice the highly oscillatory angular rate of the baseline system.**

inconsistency is due to the high attitude rates that come from the oscillatory response (visible in Figure 6.16). Quickness, therefore, is not meaningful for

Table 6.6. **Attitude Quickness Results. The Level 1 quickness requirement for target acquisition and tracking tasks is 1.2 sec^{-1} and 1.75 sec^{-1} for the roll and pitch axis, respectively.**

	roll		pitch	
	$\frac{p_{pk}}{\Delta\phi}$ sec^{-1}	ϕ_{max} deg	$\frac{q_{pk}}{\Delta\theta}$ sec^{-1}	θ_{max} deg
baseline	1.7	20	1.2	20.4
compensated	1	20.1	0.7	20.7
tendency was already slow	1.3	17.4	1.4	16.2
medium	1.7	16.6	1.7	16.2
fast	2.5	16.3	1.9	16.1

a system with oscillatory tendencies. The compensated design does not have an oscillatory response and gives results that are representative of the baseline systems performance.

The results obtained by the medium and fast designs are within the Level 1 boundaries of the rotorcraft flying qualities requirements. For a minimal attitude change of 20 degrees, the quickness requirement for target acquisition and tracking tasks is 1.2 sec^{-1} and 1.75 sec^{-1} for, respectively, the roll and pitch axes.

Same as for the bandwidth, the values used for manned requirements might not apply to unmanned operation. Based on the effect of scaling, the quickness capabilities of a small-scale rotorcraft will increase naturally with a decrease in the vehicle size. For example, based on the effect of scale on rate sensitivity (see Table 5.4), quickness is expected to increase by a factor $\sqrt{N}$ for a dynamically similar helicopter.

Achievable Attitude and Angular Rates

For large amplitude attitude changes, the flying qualities are determined by the achievable attitude angle and angular rate. For the evaluation of our controllers, we applied step attitude commands to our attitude control system and determined the largest attitude commands that can be implemented without reaching actuator saturation (±100% travel and ±800%/sec rate limit). The maximal angular rates correspond to the rates attained with these commands. The results for this test are listed in Table 6.8.

Table 6.7. **ADS-33D Requirements for large amplitude attitude changes**

	Angular Rates deg/sec		Attitude Angles degree	
	p_{max}	q_{max}	ϕ_{max}	θ_{max}
Moderate Maneuvering	±50	±13	±60	−30 +20
Aggressive Maneuvering	±50	±30	±60	±30

Table 6.8. **Maximal attitude and angular rates attained with the different controllers**

	θ_{max}	ϕ_{max}	$\delta_{lon,max}$	$\delta_{lat,max}$	q_{max}	p_{max}
baseline	25	26	10	10.4	42.1	58.8
compensated	35	35	13.7	13.5	34.7	45
slow	31	28	52.3	21.6	47.2	42.6
medium	30	21	57.8	25.8	55.5	38.9
fast	30	15	60.5	24.1	61.3	41.1

Note that the limits attained during the test are conservative, because a pure step command drives the controller very aggressively, and thus saturates the actuators. prematurely Larger commands, which drive the angular rates higher, could be attained by shaping the command input.

For the longitudinal axis, the performance increases consistently with the faster designs. For the lateral axis, the performance achieved by the different controllers, in terms of achievable attitude and angular rate, is not consistent with the designs. The issue is that, for a more aggressive controller, the actuator has a tendency to saturate for a smaller command magnitude, and a smaller command magnitude produces a smaller angular rate.

The conclusion from this test is that the design of the outer loop control (input shaping) is critical to exploit the inherent control performance for large amplitude changes.

For comparison, the large amplitude attitude requirements from ADS-33D are listed in Table 6.7. The values indicated in the table are for aggressive maneuvering. Based on the scaling laws (Table 5.4), the maximal angular rate should be about $\sqrt{N}$ times higher for a dynamically similar helicopter. Based on the scaling effect, we will need to rethink the definition of what constitutes an aggressive maneuver for small-scale rotorcraft. Thus, we will also have to redefine the boundaries that define moderate and aggressive maneuvering. For example, the X-Cell can attain roll rates up to 200 deg/sec. A computer controlled helicopter, without a stabilizer bar, could attain even higher values.

5. Conclusion

In this chapter we saw how the material form previous chapters find a direct application in the control design an analysis problems. We will briefly outline the key results and end with a note on choosing a control methodology.

Control Analysis and System Identification

With our closed-loop identification, we demonstrated that our model of the R-50 can accurately predict the performance of a closed-loop system. Good agreement between flight and model was attained in a frequency range from about 1 rad/sec to beyond 20 rad/sec. According to [85], a model should be accurate in a range of 0.3 to 3 times the intended crossover frequency. This means the R-50 model would support design up to about 9 rad/sec in crossover frequency, which is considerably higher than the 1.3 rad/sec attained with the baseline PD controller in the roll axis.

The practical value of the model was demonstrated by using it to analyze the performance of a classical attitude control system. Basic phase and gain margin stability analysis showed that the lightly damped R-F mode limits the bandwidth. The identified model helped us design a notch filter compensation based directly on the physical characteristics that were extracted from the identification. This compensation was used to enhance the baseline PD controller and allowed a substantial increase in gain margins without significant loss of phase margin. Closed-loop identification was used again and validated the effectiveness of that compensation. These results are a good illustration of the type of insight that can be obtained through the combination of system identification and control analysis.

RUAV Flying Qualities

The wide range of modalities under which unmanned vehicles could be operated will require detailed studies to determine the flying qualities requirements.

These will have to cover a range going from handling qualities-type specifications to pure task performance specifications.

The scaling effects and the generally larger freedom in the design of unmanned vehicles will result in a broad range of vehicle behaviors and capabilities. This was already demonstrated in the last chapter in the comparison between the R-50 and the X-Cell. This extension of the traditional operational envelope will probably require the definition of vehicle classes.

The traditional flying qualities framework is not directly applicable for unmanned rotorcraft. However, it provides most of the questions and criteria that will need to be explored.

Choice of Control Design Methodologies

In regard to the question of control design methodology, it is important to distinguish between the controller architecture and the actual synthesis method. Modern multivariable control design usually combines the optimization of the controller architecture and its parameters. This is the case for H_∞ method which can explicitly account for all the dynamics included in the linear design model, and can, if properly posed, achieve close to perfect control (it is still subject to the fundamental limitations of linear feedback control, e.g. effects of non-minimum phase dynamics, and saturation constraints [18]). Classical control usually refers to PID-type feedback systems and usually does not include the notion of optimization. Control design frameworks like CONDUIT [22] (for CONtrol Design Unified InTerface) allow the optimization of a user-defined control architecture against a variety of performance and stability specifications. Therefore multivariable control design can also be performed with classical architectures. However, the controller architecture should be carefully selected. For example, for systems with high-order dynamics, a PID type controller, may lack the elements that are necessary to precisely shape the close-loop characteristics. A good understanding of the vehicle dynamics can help define a controller architecture. Once such an architecture is defined, a variety of optimization methods could be used, including CONDUIT and H_∞.

Chapter 7

RESULTS, MILESTONES AND FUTURE DIRECTIONS IN AERIAL ROBOTICS

Since the completion of the identification work, some key milestones have been demonstrated in-flight with miniature unmanned rotorcraft. These results incorporate much of the knowledge presented in this book and even go beyond by covering part of the nonlinear dynamic range. From this new state of the art, the next level development necessary for advanced operations will be able to take place.

The following discussion gives a summary of the book's material with some final comments and conclusions. Then, gives an overview of the key milestones that were achieved. And finally, it outlines the problems that are driving today's research and lay out the key research directions that will make aerial robotics a research field in its own right bringing forth the full potential of autonomous flight.

1. Summary

Our original goal of an accurate model for high-bandwidth control design has yielded a significant body of knowledge about small-scale rotorcraft. In the following, I will highlight the main lessons learned.

1.1 Identification Modeling

During Carnegie Mellon's modeling-identification work on the Yamaha R-50, we collected flight data at both hover and cruise flight and estimated frequency responses for the respective flight conditions. We developed a linear parameterized model based on the fuselage-rotor equations used in full-scale helicopters; the model was extended to include the stabilizer bar, which is often used in these small-scale vehicles. The model's parameters were identified for both conditions, and its accuracy was successfully verified by a comparison of the model-predicted responses with independent flight data.

The primary result of this research is the development of an accurate model for small-scale rotorcraft. In completing this development we:

- determined what elements play a critical role in the dynamics of miniature rotorcraft
- developed a parameterized model that focuses on these essential components.

This first comprehensive application of system identification to a small-scale helicopter (10 ft rotor diameter) established that linear system identification can be successfully applied to helicopters of small size. We were able to conclude that for our vehicle, the combined fuselage-rotor-stabilizer dynamics are of primary importance and need to be taken into account. The end result is a parameterized model allowing

- high-bandwidth accuracy adequate for high-bandwidth control
- modeling of both hover an cruise conditions with an almost identical model structure; fitting the linear parameter varying (LPV) control framework

In terms of modeling, we found the following to be the case for miniature rotorcraft. Compared with full-scale rotorcraft, a simpler model structure with fewer model parameters allows a good fit of the flight data. For example, the R-50's attitude dynamics have a distinct effective third-order nature, which can be accurately fitted by the third-order fuselage-rotor-stabilizer model. Such characteristics, combined with a good understanding of the primary dynamics, allow a high degree of precision in the identification. Compared with previously existing RUAV models, our model is physically more meaningful. Both its structure and its identified model parameters have a well-defined physical meaning. For example all the key rotor, stabilizer, and rotor-fuselage coupling derivatives are perfectly defined and can be verified by helicopter theory.

The parameterized state-space model has:

- simple block structure
- rotor modeled through first-order tip-path-plane dynamics
- stabilizer bar as a secondary rotor

- cyclic control derivatives enter rotor equations of motion only
- stabilizer bar couples to the rotor through physically meaningful mixing model
- rotor forces and moments are expressed through the rotor flapping states

Finally, the high fidelity attained with a relatively simple parameterized model (compared with a highly detailed physical model) indicates that the dynamics of small-scale helicopters could be subject to less complex second-degree effects found in full-scale rotorcraft (e.g., fuselage aerodynamic effects, high-order rotor dynamics).

The parameterized model appears to be general enough to be used for modeling and identification of other small-scale rotorcraft, this was later validated by the successful identification of MIT's X-Cell helicopter. These model properties also make it useful tool for the analysis of basic flying characteristics of small-scale rotorcraft dynamics.

1.2 Characteristics

To understand the effects of size on helicopters' dynamic characteristics, we applied Froude (dynamics similarity) and Mach (rotor speed similarity) scaling rules to key flying qualities parameters. The results showed that, as size is reduced (scale factor N means 1/N the rotor diameter), attitude dynamics become faster and more sensitive to controls and disturbances:

- Froude: attitude rate sensitivity increases proportionally to $\sqrt{N}$
- Mach: attitude rate sensitivity increases proportionally to N

The thrust-to-weight ratio follows the following trend:

- Froude: thrust-to-weight ratio is unchanged
- Mach: thrust-to-weight ratio increases proportionally to N

Note also that the disc loading decreases with the vehicle size, resulting in decreasing induced power losses (factor $\sqrt{N}$). Hence a smaller vehicle will tend to be more efficient (less induced losses).

We also compared key flying and configuration parameters of small- and full-scale rotorcraft to determine how well they relate through the scaling hypotheses. We found:

- rotorcraft can be divided into two (overlapping) classes:

  - those related by Froude (dynamic) similarity, e.g. small-scale R-50 and full-scale Bell UH-1H helicopter
  - those related by Mach (speed) similarity, e.g. X-Cell and R-50

- determined the design features that most affect the vehicle performance

- demonstrated how the stabilizer bar compensates for some of the scaling effects

From the scaling results we can conclude that the Yamaha R-50 should have a mode of operation close to that of a traditional helicopter. The R-50 is clearly not designed for aggressive flight; for example, its teetering rotor head cannot produce the large control moments that are required for low-g maneuvers (pushovers, loops, inverted flight). Moreover, Carnegie Mellon's R-50 is also handicapped by its heavy instrumentation (50% of airframe weight). On the other hand, the results show that the X-Cell escapes this class of dynamically similar vehicles. It relates best to the R-50 via Mach scaling rules. This means that the X-Cell belongs to a class of more aggressive vehicle. These results give a theoretical basis for understanding the effects of scaling and validates vehicle behavior as observed during flight testing.

1.3 Control Analysis

Using the identified model for the analysis of a classical PD attitude control system, we were able to see that the stabilizer bar is largely responsible for the bandwidth limitation affecting such classical control systems. Based on our understanding that the stabilizer bar decreases the damping of the coupled rotor-fuselage mode, we were able to design a notch filter to compensate this effect. We also implemented this compensation and were able to validate the anticipated gain in stability margins using closed-loop system identification. The roughly doubled gain margins allow the implementation of higher feedback gains yielding up to four times-higher bandwidth (as seen from the optimization with CONDUIT).

These results are important for those interested in using classical control architectures on such helicopters. However, they are not the only possible way to design a high-performance control system. In fact, the model presented here opens up a wide range of possibilities for design and the results obtained with the identification experiment clearly demonstrate the suitability of the model for high-bandwidth model-based control design.

Finally, the control analysis and basic optimization also brought us to the topic of performance requirements and specifications for miniature rotorcraft. We described a few full-scale attitude performance specifications and concluded that such specifications will be important for control optimization. The problem of flying qualities for unmanned rotorcraft will need to account for a variety of new factors, including, for example, the operator interface, control modality, level of augmentation. Even fully autonomous flight will require setting up performance specifications from the task characteristics to help design the vehicle, and develop control and guidance laws that best exploit its performance in the context of a task or mission.

2. Recent Milestones

In the following we will concentrate on results for small-scale rotorcraft that are directly related to system identification modeling work presented in this book, and have been validated through flight experiments. Both the modeling and control areas have seen a substantial leap forward.

2.1 Modeling

La Civita, *et al.* [49] combined nonlinear modeling with linear system identification to produce a high-fidelity model that can be used for control analysis and simulation. MOSCA (Modeling for flight Simulation and Control Analysis) uses a nonlinear model developed from first principles. The physical model was derived from a low complexity simulation model and was extended to include the stabilizer bar dynamics and inflow dynamics (Pitt-Peters model with Keller's wake-distortion parameter). The authors used a global optimization method to identify a subset of the model parameters from frequency responses derived from flight test experiments. The fitting process is performed via linearized models. This approach was applied on Carnegie Mellon's R-50. The model was identified using the hover and cruise frequency responses described in the present work. The nonlinear model attains a match at the two operating point (with the single set of identified parameters) that is of the same order as our individually identified linear models. This technique will be particularly useful for realistic simulation of control systems throughout the entire flight envelop.

Gavrilets, *et al.* [27] developed a highly simplified nonlinear model for MIT's small-scale X-Cell helicopter. The goal of this modeling work was to capture the dynamics that are essential to aggressive flight. The model development was conducted from the ground up, starting with the linear equations of motion and extending them to include the key nonlinear effects, such as thrust and effects occurring at high incidence flight. The main idea was to use the critical level of complexity that enabled accurate modeling of the dynamics across the largest possible range of flight conditions. To guarantee sufficient accuracy, a combination of linear frequency response identification and time domain identification was used. The latter technique was used mainly for the aerodynamic derivatives. The model accuracy was verified through comparison with flight data. Moreover, the model was also flight tested in simulation by an experienced hobby pilot, to make sure the model provides realistic handling qualities, in particular for the execution of aerobatic maneuvers. This is an essential feature of this model since its main purpose is the development of control techniques for aggressive maneuvering inspired from human strategies.

2.2 Control Design

Accurate models make it possible to use a range of model-based control design methodologies, and most importantly, they allow significantly higher control performance. Two recent applications to small-scale rotorcraft projects illustrate the possibilities and performance of model-based control design in small-scale rotorcraft.

La Civita, *et al.* [50] designed and flight tested a loop shaping H_∞ controller for CMU's Yamaha R-50 helicopter. This approach from the family of robust control design methods had already been successfully used for the Harrier aircraft [39] and a full-scale helicopter [78]. The H_∞ control methodology had also previously been tested on a rigged helicopter [93] and is now used in commercially available autopilot for small-scale helicopters [91], but no results from comprehensive flight-test implementation had been published before. The loop shaping controller was designed using the model described in [49] identified with the frequency responses from chapter 2. The flight tests, during which several coordinated maneuvers were flown, demonstrated good tracking performance up to moderate speed. This loop shaping method is also suited for scheduling with flight condition, allowing the best possible performance to be attained across a range of flight conditions. This possibility was demonstrated on the Harrier [39, 69] and is currently been flight tested on the R-50 helicopter. The example in [50] is a perfect illustration of the performance that can be attained through a rigorous control design using a model identified from frequency responses.

Gavrilets, *et al.* [28] developed a control logic for automated execution of aerobatic maneuvers. This logic is made of a combination of a trim tracking controllers, used prior to and upon exit from the maneuver, and a sequencing scheme used for the implementation of the actual maneuver. The tracking controller is designed using a linear quadratic (LQ) method. The controller gains are switched as a function of operating conditions providing good tracking performance up to a 15 m/sec forward speed. The controllers' gains were computed for six operating points between 0 and 15 m/sec. The design model for each condition were obtained through linearization of the model described in previous section. To facilitate the design and implementation, a notch filter was implemented to alleviate the lightly damped rotor-fuselage mode, and the attitude dynamics were approximated as first order [59]. For the lateral-directional operation, the controller is mechanized to behave as a heading rate controller in hover, and as a turn rate controller in cruise allowing perfectly coordinated turns (zero side slip). The control sequencing scheme used for the implementation of the maneuvers is inspired from control strategy used by human in controlling such aggressive maneuvers [25]. The details defining the sequence of pre-programmed trajectories are identified from data collected in flight and from the simulator described in the previous section [80, 25, 27]. Portions of the sequence can involve control trajectories, i.e. no feedback control, or a combination of feedforward and tracking on a subset vehicle states. The timing in the sequence is in part predetermined but can also be triggered based on the vehicle states. This architecture has been successfully implemented and flight-test demonstrated for a snap roll [26]. It is currently being flight-tested for other maneuvers.

These two examples cover two primary operating modalities: tracking trajectories near trim conditions, and executing aggressive large magnitude maneuvers between trim conditions. In the following we will see how these results will lead to the flight control capabilities required for guidance and eventually autonomous operation with such agile vehicles.

3. Future Directions in Aerial Robotics

Now that the basic capabilities allowing tracking trajectories and executing maneuvers have been demonstrated, a new set of challenges will need to be addressed. The next set of problems are in large part driven by two factors: the tremendous dynamic potential of miniature rotorcraft, and the requirements set by autonomous flight.

3.1 Driving Forces

The ultimate goal of an aerial robotics system is to be able to accomplish a task without human involvement, which is summarized in one word: autonomy. Obviously there are many degrees of autonomy. These degrees can be described in terms of the amount of structure that can be assumed in the environment, and how abstract the definition of the task is.

An example task could be to build a map of an agricultural field. This operation could be executed from a pre-programmed script containing information about the precise location of the field, information about the geographical features, slope, presence of natural or man-made obstacles, actual path that needs to be flown to cover the entire area. It could however also be executed with a minimum amount of *a priori* information. Then there are the questions of performance (e.g. speed, costs and quality). This example is among the simpler applications for robotic aerial vehicles. One can imagine the surge in complexity in more hostile and dynamic environments, and for more complex tasks.

We are interested specifically in the dynamics and control challenges leading to fully autonomous operation of highly maneuverable aerial vehicle. To be able to determine how far we are from such a goal, and at the same time, understand what critical developments need to be made, we look at the capabilities that such systems will require. The problems involved in, and solutions enabling autonomy can often be ranked hierarchically, from so-called low-level problems that have a well-defined physical characteristics and often have explicit solutions, to high-level problems that are more abstract and ambiguous in their formulation and solution. We can use this framework to order the problems and task related to the dynamics and control challenges.

Compensation improves the dynamic characteristic of the vehicle to facilitate the regulation or tracking problem. In our context, the notch filter was a form of compensation.

Regulation and Tracking takes care of the operation along or near trim trajectories, as well as rejecting disturbances (e.g. wind gusts) and making up for modeling uncertainties.

Maneuvering takes care of the operation between trim conditions. Its biggest role is to incorporate high-performance maneuvers in the "collection" of vehicle behaviors.

Guidance generates feasible trajectories for the vehicle accounting for the physical environment and the task elements

Mission planning transforms/breaks down a high-level mission goal or task into elements that can be implemented by the guidance system.

Based on the milestones described in previous section, we see that the lower level problems up to the maneuvering can be considered well under way of being solved. The area of aggressive maneuvering can be considered as one of the new areas of flight control that will be critical for the broad operational capabilities.

Developments can be represented as following a bottom-up and top-down approach: the bottom-up referring to the developments going from the physical layer up to the guidance; the top-down referring to the developments taking place above guidance. The meeting point is guidance, making it simultaneously the cornerstone and the confluence point of the technological developments that need to happen in aerial robotics. The levels beyond guidance will not be discussed since they are related to a much lesser degree to the vehicle and its dynamics. Aggressive maneuvering on the other hand is the problem that incorporates the sum total of aspects related to flight dynamics: vehicle design, modeling, and control design.

3.2 Aggressive Maneuvering

Small-scale helicopters like the X-Cell can perform maneuvers that are mind-boggling. For example, they can go from hover to inverted hover without having to go through cruise flight; this is achieved through a flip maneuver. Most such maneuvers are characterized by the fact that they go through non-equilibria operating conditions; span a broad range of vehicle state variable; and have controls often reaching saturation. For example, in a roll maneuver, the vehicle will temporarily be at an attitude where no lift can be produced, its roll angle will go through the entire 360 degrees range, and the lateral cyclic input will follow a bang-bang profile.

There is a fundamental difference between the quasi-steady tracking of trajectories and maneuvering. This is a notion that is already well established among pilots (and is not limited to air vehicles); flying qualities requirements distinguish between these two domains separately. In terms of control the two domains are also distinctly set apart, as shown by the following characteristics:

Tracking operation takes place around equilibria conditions; the dynamics are mostly linear; the control problem is best addressed using continuous feedback laws.

maneuvering operation takes place between equilibria conditions; the dynamics are mostly nonlinear as the maneuver spans a range flight conditions; the system is driven mainly through feedforward type of control action; exploits full range of input and states; continuous feedback plays a secondary role.

Maneuvering can be posed as a nonlinear control problem. However, traditional nonlinear control methods (e.g. feedback linearization) have not yet been successfully demonstrated. There is a strong interest for alternative approaches. One such direction is human pilot inspired techniques, such as the one already presented in [26], which stems from the study of flight data from piloted flight experiments presented in [71, 25]. Maneuvering is the control area which needs to be addressed.

3.3 Guidance

Guidance, the problem of generating a trajectory that will guide the vehicle between locations, plays a key role in the operation of autonomous vehicles. One way to think about the boundary between control and guidance is that the control problem makes performance available and guidance exploits it. Poor or conservative guidance equates to a waste of the available assets.

The problem of guidance for highly agile vehicles in dynamics environment is in fact quite new. Missiles have long dominated this area, however, small-scale rotorcraft have redefined the problem due to the broad range of dynamic behaviors and the complexity of the environment in which they can be deployed (e.g. urban environments and even inside buildings).

One interesting and promising technique for highly agile vehicles is based on a so-called maneuver automaton (MA), introduced by Frazzoli *et al.* [24]. In a MA representation, the vehicle motion is described by a collection of motion primitives. The motion primitives are divided into trim trajectories and maneuvers. The trim trajectories correspond to relative equilibria states of the system, while maneuvers are defined as finite duration transitions between the trim states. With this intuitive representation of the vehicle dynamics, the trajectory between two positions and vehicle states is found by searching the sequence of motion primitives which will best satisfy an objective function.

This is a sequential decision problem which Frazzoli *et al.* showed could be elegantly posed as a dynamic program, and solved off-line allowing real-time implementation.

The advantage of such a method, besides computational tractability, is that it can account for any maneuver as long as it is feasible. Hence it can potentially exploit the full dynamic potential of the vehicle. Such a motion planning system was developed for the X-Cell and implemented in simulation [62]. The preliminary results are very encouraging but have highlighted a series of issues that will need to be addressed for an efficient implementation.

3.4 Final Note on Rotorcraft

Small-scale helicopters like the X-Cell demonstrate that traditional rotorcraft are highly capable platforms. The flapping rotor can produce large control forces and moments and also produce high thrust-to-weight ratios. The former results in high rate sensitivity which allows for quick changes in thrust and travel direction; the latter can literally cancel the effect of gravity. Scaling effects, if combined with proper design features (e.g. no stabilizer bar, stiff rotor head, fast rotor) will result in a new generation of rotorcraft with dramatically higher performance than traditional rotorcraft and even today's most agile RC helicopters (which still have to satisfy the human pilot's constraints).

The making of such an aerial robot will naturally put stringent demands on other technologies (sensing, communication, information processing), and motivate further engineering problems. There will be exciting research opportunities in a variety of areas, making the aerial robot automaton a centerpiece of modern robotics.

References

[1] General specification for flight control systems. Technical Report MIL-F-9490D, United States Air Force, St. Louis, Missouri, 1975.

[2] Handling qualities requirements for military rotorcrafts. Technical Report ADS-33D-PRF, United States Army and Troop Command, Aviation Research and Development Center, St. Louis, Missouri, 1996.

[3] O. Amidi, T. Kanade, and R. Miller. "Vision-based Autonomous Helicopter Research at Carnegie Mellon Robotics Institute". Japan, April 1998. Proceedings of Heli Japan'98.

[4] S. Atkins, W. Hall, P. DeBitetto, and D. Cohn. "The MIT / Draper Laboratory Autonomous Helicopter". Technical report, Massachusetts Institute of Technology, Draper Laboratory, 1995.

[5] J.S. Bendat and A.G. Piersol. *"Engineering Applications of Correlation and Spectral Analysis"*. John Wiley & Sons, New York, NY, 1993.

[6] A.R.S. Bramwell, G. Done, and D. Balmford. *"Bramwell's Helicopter Dynamics"*. Butterworth Heinemann, Oxford, UK, 2001.

[7] P.B. Bruce, J.E.F. Silva, and M.G. Kellett. "Maximum Likelihood Identification of a Rotary-Wing RPV Simulation Model From Flight-Test Data". Number 98-4157, Boston, MA, 1998. Proceedings of the AIAA Atmospheric Flight Mechanics Conference and Exhibit.

[8] S.M. Burk and C.F. Wilson Jr. "Radio-Controlled Model Design And Testing Techniques for Stall/Spin Evaluation of General-Aviation Aircraft". National Business Aircraft Meeting, 1975.

[9] C.-T. Chen. *Linear System Theory and Design*. Oxford University Press, Oxford, UK, 1984.

[10] R. Chen. "Effects of Primary Rotor Parameters on Flapping Dynamics". Technical Report TP-1431, NASA, 1980.

[11] R. Chen. "Simplified Rotor System Mathematical Model for Piloted Flight Dynamics Simulation". Technical Report Technical Memorandum 78575, NASA Ames Research Center, Moffett Field, CA, 1980.

[12] J.D. Colbourne. "Control Law Design and Optimization for Rotorcraft Handling Qualities Criteria Using CONDUIT". Montreal, Canada, 1999. 55th Annual Forum of the American Helicopter Society.

[13] J.D. Colbourne, L.S. Cicolani, C.R. Frost, M.B. Tischler, C. Tomashofski, and T. LaMontagne. "System Identification and Control System Design for the BURRO Autonomous UAV". Virginia Beach, VA, May 2000. 56th Annual Forum of the American Helicopter Society.

[14] J.D. Colbourne, M.B. Tischler, and K. Rodgers. "Flight Control Design for an Unmanned Rotorcraft Program with a Rapid Development Schedule". Washington, DC, May 2001. 57th Annual Forum of the American Helicopter Society.

[15] G.E. Cooper and R.P. Harper. "The Use of Pilot Ratings in the Evaluation of Aircraft Handling Qualities". Technical Report TM D-5133, NASA, 1969.

[16] K.J. Corfeld, R.C. Strawn, and L.N. Long. "Computational Analysis of a Prototype Martian Rotorcraft Experiment". Big Sky, MT, March 2002. IEEE Aerospace Conference.

[17] H. C. Curtiss, W.F. Putman, and E. Martinez. "The Evaluation of Stability and Control Characteristics of Aircraft at Low Speed Using Dynamically Similar Models in Semi-Free Flight". Washington, D.C., May 1962. 18th Annual Forum of the American Helicopter Society.

[18] J. Doyle, B. Francis, and A. Tannenbaum. *Feedback Control Theory*. Macmillan, New York, NY, 1992.

[19] E. Feron *et al.* "The MIT Entry into the 1998 AUVS International Aerial Robotics Competition". Association for Unmanned Vehicle Systems International (AUVSI), June 1998.

[20] J.-M. Biannic et al. "Parameter Varying Control of a High-Performance Aircraft". *Journal of Guidance, Control, and Dynamics*, 20(2):225–231, 1997.

[21] M. Sugeno et al. "Fuzzy Hierarchical Control of an Unmanned Helicopter". Technical report, Tokyo Institute of Technology, Tokyo, Japan, 1993.

[22] M.B. Tischler et al. "A Multidisciplinary Flight Control Development Environment and Its Application to a Helicopter". *IEEE Control Systems Magazine*, 19(4):22–33, August 1999.

[23] J.W. Fletcher. "Identification of UH-60 Stability Derivative Models in Hover from Flight Test Data". *Journal of the American Helicopter Society*, 1995.

[24] E. Frazzoli, M.A. Dahleh, and E. Feron. Real-time motion planning for agile autonomous vehicles. *AIAA Journal of Guidance, Control, and Dynamics*, 25(1):116–129, 2002.

[25] V. Gavrilets, E. Frazzoli, B. Mettler, M. Piedmonte, and E. Feron. "Aggressive Maneuvering of Small Autonomous Helicopters: A Human-Centered Approach". *International Journal of Robotics Research*, pages 795–807, October 2001.

[26] V. Gavrilets, I. Martinos, B. Mettler, and E. Feron. "Aggressive Maneuvering Flight Tests of a Miniature Robotic Helicopter". Sant'Angelo d'Ischia, Italy, July 2002. Proceedings of the 8th International Symposium on Experimental Robotics.

[27] V. Gavrilets, B. Mettler, and E. Feron. "Nonlinear Model for a Small-Size Acrobatic Helicopter". Number AIAA 2001-4333, Montreal, Canada, August 2001. Proceedings of the American Institute of Aeronautics Guidance, Navigation, and Control Conference.

[28] V. Gavrilets, B. Mettler, and E. Feron. "Control Logic for Automated Aerobatic Flight of a Miniature Helicopter". Monterrey, CA, August 2002. Proceedings of the American Institute of Aeronautics Guidance, Navigation, and Control Conference.

[29] A. Gessow and G.C.J. Myers. *Aerodynamics of the Helicopter.* Frederick Ungar Publishing Co., New York, NY, 1952.

[30] J.A. Ham, G.K. Gardner, and M.B. Tischler. "Flight-Testing and Frequency-Domain Analysis for Rotorcraft Handling Qualities". *Journal of the American Helicopter Society*, April 1995.

[31] P.G. Hamel and R.V. Jategaonkar. "The Evolution of Flight Vehicle System Identification". *Agard Lecture Series on Rotorcraft System Identification*, 1995.

[32] P.G. Hamel and J. Kaletka. "Advances in Rotorcraft System Identification". *Progress in Aerospace Sciences*, Vol. 33:259–284, 1997.

[33] R.S. Hansen. "Toward a Better Understanding of Helicopter Stability Derivatives". *Journal of the American Helicopter Society*, 29(1):15–24, 1982.

[34] R.K. Heffley. "A Compilation and Analysis of Helicopter Handling Qualities Data; Volume I: Data Compilation". Technical Report CR-3144, NASA, 1979.

[35] R.K. Heffley. "A Compilation and Analysis of Helicopter Handling Qualities Data; Volume II: Data Analysis". Technical Report CR-3145, NASA, 1979.

[36] R.K. Heffley, S.M. Bourne, H.C. Curtiss, W.S. Hindson, and R.A. Hess. "Study of Helicopter Roll Control Effectiveness Criteria". Technical Report TM-85990, NASA, 1984.

[37] R.K. Heffley and M.A. Mnich. "Minimum-Complexity Helicopter Simulation Math Model". Technical Report USAAVSCOM Technical Report 87-A-7, NASA, Moffett Field, CA, 1987.

[38] Kathryn B. Hilbert. "A Mathematical Model of the UH-60 Helicopter". Technical Report NASA TM-85890, NASA, Moffett Field, CA, 1984.

[39] R. Hyde. *H_∞ Aerospace Control Design - A VSTOL Flight Application*. Springer Verlag, 1995.

[40] R.A. Hyde and K. Glover. "Meeting VSTOL Aircraft Performance Requirements using Scheduled H_∞ Controllers". *Proceedings of the 1991 American Control Conference*, 1991.

[41] L.B. Jackson. *Signals, Systems, and Transforms*. Addison-Wesley, 1991.

[42] W. Johnson. *Helicopter Theory*. Princeton University Press, Princeton, NJ, 1980.

[43] C. A. Robinson Jr. "Amphibious Warfare Changes Prompt Entrepreneurial Ways". Fairfax, VA, December 1998. Armed Forces Communications and Electronic Association.

[44] J. Kaletka and O. Rix. "Aspects of System Identification of Helicopters". Technical report, Braunschweig, Germany, 1978.

[45] S. Kannan, A.D. Kahn, and I. Yavrucuk. "GTMARS - Flight Controls and Computer Architecture". Technical report, School of Aerospace Engineerng, Georgia Institute of Technology, Atlanta, GA, 2000.

[46] M.G. Kellet. "Scheduled Robust Control for Helicopter Applications". Singapore, 1992. ICARCV'92. Second International Conference on Automation, Robotics and Computer Vision.

[47] A.K. Kim and D.M. Tilbury. "Mathematical Modeling and Experimental Identification of a Model Helicopter". AIAA, 1998.

[48] T.J. Koo and S. Sastry. "Output Tracking Control Design of a Helicopter Model Based on Approximate Linearization". Tampa, FL, 1998. Proceedings of the 37th IEEE Conference on Decision and Control.

[49] M. LaCivita, W.C. Messner, and T. Kanade. "Modeling of Small-Scale Helicopters with Integrated First-Principles and System Identification Techniques". Montreal, Canada, June 2002. Proceedings of the 58th Forum of the American Helicopter Society.

[50] M. LaCivita, G. Papageorgious, W.C. Messner, and T. Kanade. "Design and Flight Testing of a High-Bandwidth H_∞ Loop Shaping Controller for a Robotic Helicopter". Number AIAA-2002-4836, Monterrey, CA, August 2002. AIAA Guidance Navigation and Control Conference.

[51] H.L. Langhaar. *Dimensional Analysis and Theory of Models*. John Wiley & Sons, New York, 1951.

[52] J.G. Leishman. *Principles of Helicopter Aerodynamics*. Cambridge University Press, Cambridge, UK, 2000.

[53] L. Lipera, J.D. Colbourne, M.B. Tischler, M.H. Mansur, M.C. Rotkowitz, and P. Patangui. "The Mircro Craft iSTAR Micro Air Vehicle: Control System Design and Testing". Washington, DC, May 2001. 57th Annual Forum of the American Helicopter Society.

[54] L. Ljung. *System Identification*. Prentice-Hall, 1987.

[55] The Math Works Inc. ed. T.M.W. Inc., Natick, MA. *User's Guides for MATLAB, SIMULINK, Toolboxes*, 2001.

[56] D.T. McRuer, I. Ashkenas, and D. Graham. *Aircraft Dynamics and Automatic Control*. Princeton University Press, Princeton, NJ, 1973.

[57] D.T. McRuer and E.S. Krendel. "Mathematical Models of Human Pilot Behavior". Technical Report AGARD-AG-188, AGARD, 1974.

[58] B. Mettler, C. Dever, and E. Feron. "Identification Modeling, Flying Qualities and Dynamic Scaling of Miniature Rotorcraft". Berlin, Germany, May 2002. NATO SCI Symposium on Challenges in Dynamics, System Identification, Control and Handling Qualities For Land, Air, Sea and Space Vehicles.

[59] B. Mettler, V. Gavrilets, E. Feron, and T. Kanade. "Flight-Test Evaluation of Dynamic Compensation for High-Bandwidth Control of Small-Scale Helicopter". San Francisco, CA, Jan 2002. American Helicopter Society Test and Evaluation.

[60] B. Mettler, M.B. Tischler, and T. Kanade. "System Identification of Small-Size Unmanned Helicopter Dynamics". Montreal, Canada, May 1999. Proceedings of the 55th Forum of the American Helicopter Society.

[61] B. Mettler, M.B. Tischler, and T. Kanade. "System Identification Modeling of a Small-Scale Unmanned Rotorcraft for Control Design". *Journal of the American Helicopter Society*, 47(1):50–63, January 2002.

[62] B. Mettler, M. Valenti, T. Schouwenaars, E. Frazzoli, and E. Feron. "Rotorcraft Motion Planing for Agile Maneuvering". Montreal, Canada, June 2002. Proceedings of the 58th Forum of the American Helicopter Society.

[63] R. Miller, B. Mettler, and O. Amidi. "Carnegie Mellon University's 1997 International Aerial Robotics Competition Entry". Orlando, FL, June 1997. Association for Unmanned Vehicle Systems International (AUVSI).

[64] F.C. Moon. *Applied Dynamics*. John Wiley & Sons, Inc., 1998.

[65] J. Morris, M.V. Nieuwstadt, and P. Bendotti. "Identification and Control of a Model Helicopter in Hover". Proceedings of the American Control Conference, June 1994.

[66] A.V. Oppenheim and R.W. Schafer. *Discrete-Time Signal Processing*. Signal Processing Series. Prentice Hall, Englewood Cliffs, 1989.

[67] G.D. Padfield. "The Influence of Flying Qualities on Operational Agility". AGARD Meeting on Technology for Highly Manoeuvrable Aircraft, October 1993.

[68] G.D. Padfield. *Helicopter Flight Dynamics: The Theory and Application of Flying Qualities and Simulation Modeling.* AIAA Education Series, Washington, 1996.

[69] G. Papageorgious and K. Glover. "Taking Robust LPV Control Into Flight on the VAAC Harrier". pages 4559–4564. Proceedings of the 39th IEEE Conference on Decision and Control, 2000.

[70] M.E. Peters and D. Andrisani. "The Determination of Longitudinal Flying Qualities Requirements for Light Weight Unmanned Aircraft". New Orleans, 1997. AIAA Guidance Navigation, and Control Conference.

[71] M. Piedmonte and E. Feron. "Aggressive Maneuvering of Autonoumous Aerial Vehicles: A Human-Centered Approach". Snowbird, UT, October 1999. International Symposium on Robotics Research.

[72] J.V.R. Prasad, A.J. Calise, and J.E. Corban. "Implementation of Adaptive Nonlinear Controller for Flight Test on an Unmanned Helicopter". Proceeding of the IEEE Conference on Decision and Control, 1998.

[73] J.V.R. Prasad, A.J. Calise, J.E. Corban, and Y. Pei. "Adaptive Nonlinear Controller Synthesis and Flight Test Evaluation on an Unmanned Helicopter". IEEE International Conference on Control Applications, 1999.

[74] R.W. Prouty. *Helicopter Performance, Stability and Control.* Krieger Publishing Company, Malabar, FL, 1995.

[75] D.H. Shim, H.J. Kim, and S. Sastry. "Control System Design for Rotorcraft-Based Unmanned Aerial Vehicles using Time-domain System Identification". Anchorage, 2000. IEEE International Conference on Control Applications.

[76] S. Skogestad and I. Postlethwaite. *Multivariabl Feedback Control, Analysis and Design.* John Wiley & Sons, New York, NY, 1996.

[77] J.-J. Slotine and W. Li. *Applied Nonlinear Control.* Prentice Hall, Englewood Cliffs, NJ, 1991.

[78] A. Smerlas, I. Postlethwaite, D. Walker, M. Strange, J. Howitt, R. Norton, A. Gubbels, and S. Baillie. "Design and Flight Testing of an H_∞ Controller for the NRC Bell 205 Experimental Fly-By-Wire Helicopter". Number AIAA 98-4300. Proceedings of the AIAA Guidance, Navigation, and Control Conference, 1998.

[79] S.A. Snell, D.F. Enns, and W.L. Garrard. "Nonlinear Inversion Flight Control for a Supermaneuverable Aircraft". *Journal of Guidance, Control and Dynamics*, 15(4):976–984, 1992.

[80] K. Sprague, V. Gavrilets, D. Dugail, B. Mettler, and E. Feron. "Design and Applications of an Avionics System for a Miniature Acrobatic Helicopter". Daytona Beach, FL, 2001. AIAA Digital Avionics Systems Conference.

[81] B. Stevens and F. Lewis. *Aircraft Control and Simulation.* John Wiley & Sons Inc., 1992.

[82] M.D. Takahashi, J.W. Fletcher, and M.B. Tischler. "Development of a Model Following Control Law for Inflight Simulation using Analytical and Identified Models". Fort Worth, TX, 1995. 51st Forum of the American Helicopter Society.

[83] P.D. Talbot, B.E. Tinling, W.A. Decker, and R.T.N. Chen. "A Mathematical Model of a Single Main Rotor Helicopter for Piloted Simulation". Technical Report NASA TM-84281, NASA, Moffett Field, CA, 1982.

[84] M.B. Tischler. "Frequency-Response Identification of XV-15 Tilt-Rotor Aircraft Dynamics". Technical Report TM-89428, NASA, Moffett Field, CA, 1987.

[85] M.B. Tischler. "System Identification Requirements for High-Bandwidth Rotorcraft Flight Control System Design". *Journal of Guidance, Control Dynamics*, 13(5):835–841, 1990.

[86] M.B. Tischler, editor. *System Identification Methods for Aircraft Flight Control Development and Validation*, pages 35–69. Taylor & Francis, London, 1996.

[87] M.B. Tischler and M.G. Cauffman. "Frequency-Response Method for Rotorcraft System Identification: Flight Application to BO-105 Coupled Rotor/Fuselage Dynamics". *Journal of the American Helicopter Society*, 37(3):3–17, 1992.

[88] M.B. Tischler and M.G. Cauffman. *"Comprehensive Identification from Frequency Responses; An interactive facility for system identification and verification, Class Notes and User's Manual"*. NASA Ames Research Center, Moffett Field, CA, 1994.

[89] C.A. Tomashofski and M.B. Tischler. "Flight Test Identification of SH-2G Dynamics in Support of Digital Flight Control System Development". Montreal, Canada, 1999. Proceedings of the 55th Forum of the American Helicopter Society Proceedings.

[90] D.J. Walker and I. Postlethwaite. "Advanced Helicopter Flight Control Using Two-Degree-of-Freedom H_∞ Optimization". *Journal of Guidance, Control, and Dynamics*, 19(2), March-April 1996.

[91] weControl. "Data Sheet of the wePilot1000 Flight Control System for Small Remote-Controlled Helicopters". Technical report, weControl GmbH, c/o Institut für Mess- und Regeltechnik, ETH Zentrum ML, CH-8092 Zurich, 2001.

[92] M.W. Weilenmann. *Robust Mehrgrössen-Regelung eines Helicopters.* PhD thesis, Swiss Federal Institute of Technology, Zurich, Switzerland, 1994.

[93] M.W. Weilenmann, U. Christen, and H. P. Geering. "Robust Helicopter Position Control at Hover". pages 2491–2495, Baltimore, MD, June-July 1994. Proceedings of the 1994 American Control Conference.

[94] M.W. Weilenmann and H.-P. Geering. "A test Bench for the Rotorcraft Hover Control". Monterey, CA, 1993. AIAA Guidance Navigation and Control Conference.

[95] J.N. Williams, J.A. Ham, and M.B. Tischler. "Fligt Test Manual". Technical Report AQTD Project No. 93-14, U.S. Army Aviation Technical Test Center, Edwards AFB, CA, 1995.

[96] A. Yue, I. Postlethwaite, and G. Padfield. "H_∞ Design and the Improvement of Helicopter Handling Qualities". *Vertica*, 13(2), 1989.

Index

GPSR Compliance

The European Union's (EU) General Product Safety Regulation (GPSR) is a set of rules that requires consumer products to be safe and our obligations to ensure this.

If you have any concerns about our products, you can contact us on ProductSafety@springernature.com

In case Publisher is established outside the EU, the EU authorized representative is:

Springer Nature Customer Service Center GmbH
Europaplatz 3
69115 Heidelberg, Germany

Batch number: 10755292

Printed by Printforce, the Netherlands